U.S. NAVY AND MARINE CORPS CAMPAIGN AND COMMEMORATIVE MEDALS

Edward J. Emering

Schiffer Military/Aviation History
Atglen, PA

Dedication:

To my loving wife and companion, Cali and to my children: Whitney, Scott, Eric and Ian.

Acknowledgments:

To Kathy B. and Nettie E. for support and encouragement. Special thanks to Ned Broderick (National Vietnam Veterans Art Museum, Chicago, IL), E. C. Finney, Jr. (Naval Historical Center, Washington, D.C.), Timothy Frank (The Navy Museum, Washington, D.C.), W. D. Grissom, Sr., Cindee Herrick (Coast Guard Museum, New London, CT), Charles P. McDowell (Orders and Medals Society of America), Holly Reed (National Archives - Still Picture Branch) and George M. Shank.

Book Design by Ian Robertson.

Printed in China.
ISBN: 0-7643-0386-4

We are interested in hearing from authors with book ideas on related topics.

Published by Schiffer Publishing Ltd.
4880 Lower Valley Road
Atglen, PA 19310
Phone: (610) 593-1777
FAX: (610) 593-2002
E-mail: Schifferbk@aol.com
Please write for a free catalog.
This book may be purchased from the publisher.
Please include $3.95 postage.
Try your bookstore first.

INTRODUCTION

It is with immense pleasure that I have undertaken this work. As an avid collector of militaria, I started collecting the diverse campaign and commemorative medals of the U.S. Navy and Marine Corps while on active duty in the Navy during the mid 60s. As my collection grew, I began giving thought to the display of the medals. It was then that I realized that embodied in them was a lesson in U.S. history. It is a lesson that begins with our Civil War and carries through our country's global expansionist program (often translated as protecting U.S. overseas financial interests) and on into the two great World Wars. This period is followed by a series of conflicts with the forces of communism around the world, which collectively are viewed in some quarters as World War III.

Although U.S. campaign medals in general do not reflect the flare of other nations, they may generally be considered well made and finely detailed. They may also be viewed as rather recent developments when compared to those of other nations. The first five Navy campaign medals were issued less than 100 years ago in 1908 (Civil War Campaign, West Indies Campaign, Spanish Campaign,

Philippine Campaign and the China Relief Expedition Medals). The Marine Corps, with one exception, has utilized the Navy campaign medals as their own, adding a distinctive reverse (see glossary). in the early years, they were generally produced under private contract. In later years, campaign medals were produced with a distinctive "M. No." prefix by the U.S. Mint. The obverse (see glossary) of the early Navy campaign medals was shared with the Marine Corps. The reverse was distinct to each service. In most instances, campaign medals were issued with ribbons common to both the Navy and Marine Corps. Some ribbons were subsequently altered for political reasons. This category of campaign medals with altered ribbons include: the West Indies Campaign Medal, the Spanish Campaign Medal, the Philippine Campaign Medal and the China Relief Expedition Medal. The ribbon of the World War I Army of Occupation Medal, which is also awarded to Navy personnel serving with the occupation forces, was also altered, but the ribbon change occurred prior to the original issue date. The planchets (see glossary) of most campaign medals measure approximately 1.5 inches in diameter.

Distinctive Navy campaign medal reverse with curved "For Service."

Distinctive Navy campaign medal reverse with straight "For Service."

Another interesting and important aspect of the early campaign medals is rim numbering, i.e. the numbering of the medal along its rim. In the case of the Navy campaign medals, only the Philippine Campaign, Mexican Service, Second Nicaraguan Campaign and the Yangtze Service Medals bear an "M. No." prefix applied by the U.S. Mint. In the case of the Marine Corps, only the Second Nicaraguan Campaign, the Yangtze Service and the Marine Corps Expeditionary Medals bear an "M. No." prefix. All other Navy and Marine Corps numbered campaign medals through the 1930's are unprefixed and the size and style of the numbers varies by maker. After that date, the practice of numbering campaign medals was abandoned.

Until the issue of the Dominican Campaign Medal in 1921, the reverse of all Navy and Marine Corps campaign medals featured a curved "For Service" inscription. Starting with the Dominican Campaign Medal the reverse design was altered to feature a straight "For Service." This reverse design was consistently used for Navy campaign medals until WW II. While the issue of reverse design may seem trivial, an example of the China Relief Expedition Medal (1901 obverse) was found in the early 70s with the first version suspension ribbon, showing proper wear, and bearing an unprefixed number in the 200 range. Unfortunately, the reverse utilized a straight "For Service." This was clearly a FAKE! Instead of being worth in excess of $1,000.00, it was worth nothing more than the price of a curiosity piece. This is a good example of why it is important to understand all of the various aspects of authenticating an original medal.

Beginning with the American Defense Service Medal, campaign medals began being standardized for all branches of the service with reverses designed to compliment the obverse. This general practice has continued into the modern era. Examples include not only the World War II campaign and victory medals, but also the campaign medals for Korea, Vietnam and Bosnia.

Several of the commemorative medals were struck in precious metals and in extremely limited numbers. The fact that certain commemorative medals were issued in very minimal numbers makes them extremely valuable with prices easily ranging into five figures. The actual value of certain commemorative medals can only be established via the auction process in a fashion similar to the manner in which the value of a rare painting is established. In two cases, the Manila Bay Medal and the Sampson Medal, each medal was individually rim named, i.e. the recipient's name and rank was engraved on the rim. In almost every instance, they were privately manufactured and issued to make up for a lack of federally authorized medals. Often, unit commanders financed the manufacture of these medals personally. In yet other cases, Congress authorized the manufacture of specially designed medals to recognize historic deeds or events for which no federally authorized medal existed. Sometimes this was done in acquiescence to public pressure.

Presented herein is a comprehensive review of each of the Navy's campaign and significant commemorative medals along with the one unique Marine Corps campaign medal and a brief history of the events, often cataclysmic, surrounding their issue. Many well written historical treatises exist for the reader, who wants to delve deeper into the events cited herein. A few of the more important works have been cited in the bibliography. I highly recommend pursuing these references for deeper insight into the people and events that have shaped our country's military history.

Distinctive Marine Corps campaign medal reverse with curved "For Service."

Distinctive Marine Corps campaign medal reverse with straight "For Service."

CONTENTS

GLOSSARY OF TERMS

1. Atlantic Theater - A reference to battles or operations taking place in or around the Atlantic Ocean.

2. Banana War - One of a series of skirmishes fought during the early part of the 20th century in tropical settings such as Mexico, Nicaragua, Haiti, etc.

3. Brooch - The suspension bar used to hang the ribbon and affix the medal to the uniform.

4. Clasp - A device affixed to the suspension ribbon identifying a specific campaign or operation.

5. Cross Pattee - A cross with arms each of which curve outward and have a straight end.

6. Fake - An unauthorized or unofficial restrike.

7. MC - Marine Corps.

8. M. No. - The prefix used on certain Navy and Marine Corps campaign medals manufactured by the U.S. Mint.

9. MOH - Medal of Honor.

10. Obverse - The front of the planchet.

11. Pacific Theater - A reference to the battles or operations taking place in and around the Pacific Ocean.

12. Planchet - The main part of the medal as opposed to the suspension.

13. Restrike - An official reissue using either the original dies or newly created dies.

14. Reverse - The back of the planchet.

15. Service Ribbon - A ribbon worn wrapped around a narrow bar which is attached to the uniform to either represent a full size medal, which has been awarded, or to act as an award, itself, when no corresponding full size medal exists.

16. USN - U.S. Navy.

1

THE EARLY CAMPAIGN MEDALS

Civil War Campaign Medal

Instituted on June 27, 1908, this is the first campaign for which a medal was authorized for Navy personnel. The obverse of the medal features the famous March 9, 1862 sea battle between the two ironclads, the Monitor, under the command of Lieutenant John L. Warden, and the CSS Virginia (formerly the Merrimack) at Hampton Roads, Virginia. Approximately 2,700 Navy medals with non-prefixed rim numbers were issued. Only 200 non-prefixed Marine Corps medals were issued. The issue of this medal more than 40 years after the conclusion of the War actually followed that of the Spanish Campaign medal. The limited numbers struck also reflect the few eligible recipients still living to claim it. Subsequent restrikes of this medal are unnumbered.

War With Spain

The following three Navy campaign medals (West Indies, Spanish and Philippine Campaigns) were spawned by the Spanish-American War, which had its roots in our problems with the manner in which Cuba was being governed. In addition, this War spawned four commemorative medals (Dewey, Sampson, Specially Meritorious and Cardenas) which are discussed in detail in Chapter V. The cause of the War was twofold. First, American President William McKinley was displeased with the often brutal way in which Spain governed Cuba. Of particular note to the Americans was the heavy handed manner in which Spain quelled a Cuban revolutionary movement in 1895. Second was the mysterious explosion and sinking (now known to be accidental) of the Battleship U.S.S. Maine inside Havana Harbor on February 5, 1898. The War itself vindicated the need for America to have a strong naval force. Stunning victories in the Pacific and Caribbean laid the basis for America's global expansionism during the early part of the 20th century.

Navy Civil War Campaign Medal obverse.

Navy Civil War Campaign Medal reverse.

ENDORSEMENT.

Isaac K Archer
Barrington NJ

Sir: The Bureau forwards herewith a **Civil War** Campaign Badge (No. 2298), awarded by act of Congress approved May 13, 1908, directing the preparation and distribution of badges to the officers and men of the Navy and Marine Corps of the United States who participated in engagements and campaigns deemed worthy of such commemoration.

The badge is issued in recognition of services on board the U.S.S *Jamestown*

For the purpose of identification this badge is marked with a number on the rim, which is recorded. The Bureau authorizes the engraving on the rim of this badge of his name, rate at the time, and the name of the vessel on which he was serving at **his own expense.**

Please have receipt below signed and returned to the Bureau.

Very respectfully,

VICTOR BLUE
Chief of Bureau.

July 25th, 1914

BUREAU OF NAVIGATION,
Navy Department, Washington, D.C.

I have this day received the badge (No. 2298) above mentioned.

Isaac K Archer

Award document and receipt for Navy Civil War Campaign Medal, rim number 2298, awarded to Isaac K. Archer, Barrington, N.J., who served without pay on the USS Jamestown during the Civil War. Courtesy George Shank.

West Indies Campaign Medal Version I and II

This medal was authorized in 1908 for veterans of the West Indies campaign who participated in the Atlantic Theater of the Spanish American War. Although these veterans were also eligible for the Sampson Medal (see Chapter V) that medal was officially deemed to be a commemorative medal and not a campaign medal. In 1910, the West Indies Campaign Medal was replaced by the Spanish Campaign Medal (see following discussion). The original ribbon for the West Indies Campaign Medal was yellow with two wide red stripes. Since this medal was replaced by the Spanish Campaign Medal and that medal underwent a subsequent ribbon change in 1913, the West Indies Campaign Medal is usually encountered with the same replacement ribbon used for the Spanish Campaign Medal. The replacement ribbon is yellow with two wide blue stripes. Also, these two medals share a common unprefixed rim numbering sequence ranging from 1 to 6050 for Navy issues. Marine Corps numbers range from 1 to 900 on the same continuous basis for both medals. Restrikes are unnumbered.

West Indies Campaign Medal Version I.

West Indies Campaign Medal Version II.

West Indies Campaign Medal reverse.

Spanish Campaign Medal Version I and II

This is the first Navy campaign medal to be struck. Issued in 1908, it was originally authorized only for issue to Navy personnel participating in the Pacific Theater of the Spanish American War. In the 1920's, the award criteria was relaxed to provide for award of this medal to all those who served in the Navy during the Spanish American War. It in effect replaced the West Indies Campaign Medal. In 1913, the ribbon was changed as previously mentioned in the

discussion of the West Indies Campaign Medal. Some speculate that the change was made to mirror the Army's Spanish Campaign Medal, but elsewhere it has been speculated that the change from the original colors, which were coincidentally the colors of the Spanish national flag, was made so as not to embarrass Spain, in deference to the peaceful relations which had developed between our contries.

Spanish Campaign Medal Version I.

Spanish Campaign Medal Version II.

Spanish Campaign Medal reverse.

Philippine Campaign Medal Version I and II

Established on June 27, 1908, the Philippine Campaign Medal was awarded to Navy personnel serving on board 64 ships, some of which also qualified for the Manila Bay Commemorative Medal (as noted by an *), and four land bases (Cavite, Olongapo, Polloc and Isabella de Basilan) between February 4, 1899 and March 10, 1906. The medal bears the dates 1899-1903 on its obverse, which features the Gates of Manila.

After annexing the Philippines following the Spanish American War, America met with stiff resistance from a guerrilla force led by Emilio Aguinaldo. Governor-General of the Islands and future U.S. President, William Howard Taft appointed Major General Adna Chaffee to put down the insurrection after General Aguinaldo declared independence (General Chaffee was extremely influential in convincing Congress to approve the concept of issuing campaign medals). The initial battle between U.S. forces and 40,000 "insurrectos" ultimately developed into a protracted guerrilla war, lasting until July 4, 1902. American casualties were 1,000 killed in action and more than 3,000 wounded. Although unrest continued, President Theodore Roosevelt declared the conflict at an end. Still, a major presence of both Army and Navy personnel would be required in the Islands to maintain order.

The original ribbon was red with a wide yellow center stripe. In 1913 it was changed to blue with two wide red stripes. The medal was issued with unprefixed rim numbers ranging from 1 to 4192 and the traditional curved "For Service" and with "M.No." prefixed numbers ranging from 4193 to 4392 and a straight "For Service" on the reverse. Marine Corps medals were issued with unprefixed numbers ranging from 1 to 1275 and utilizing a curved "For Service" only.

The ships involved in the Philippine Campaign include:

Albany	Charleston	Manila	Paraguay
Albay	Chauncey	Manileno	Petrel*
Annapolis	Concord*	Marietta	Piscataqua
Arayat	Culgoa	Mariveles	Princeton
Arethust	Don Juan	Mindoro	Quiros
Baltimore*	De Austria	Monadnock	Rainbow
Barry	Frolic	Monterey	Samar
Basco	Gardoqui	Nanshan	Solace
Bennington	General Alva	Nashville	Urdaneta
Boston*	Glacier	Newark	Vicksburg
Brooklyn	Helena	New Orleans	Villalobos
Buffalo	Iris	New York	Wheeling
Calamianes	Isla De Cuba	Olympia*	Wompatuck
Callao	Isla De Luzon	Oregon	Yorktown
Castine	Kentucky	Pampanga	Yosemite
Celtic	Leyte	Panay	Zafiro

Philippine Campaign Medal Version I

Philippine Campaign Medal Version II.

Philippine Campaign Medal reverse.

China Relief Expedition Medal Version I and II

This campaign medal was authorized on June 27, 1908 for Navy personnel serving ashore and on contiguous bodies of water and rivers during the Boxer Rebellion. The Boxers (a name applied by Westerners to the Righteous Fists of Fury sect) were a secret, violent north Chinese sect. When this group threatened U.S. interests, our naval forces joined with those of Austria-Hungary, France, Italy, Japan, Great Britain, Germany and Russia to restore peace. The first contingent of U.S. personnel consisted of 49 Marines and six sailors under the command of Captain John T. Myers. They arrived in Peking on May 31, 1900. When the German ambassador was shot by a Chinese soldier on June 20, the allied forces prepared to defend the barricaded Legation (diplomatic) Quarter. The first Boxer attack came that very afternoon. In addition, the Boxers also besieged the diplomatic settlement at Tientsin.

A multinational force of 6,500 (1,000 Americans commanded by General A. R. Chaffee) relieved Tientsin on July 14. An even greater multinational force of 18,600 (2,500 Americans) under the command of British General Sir Alfred Gaselee advanced on Peking. Peking's Legation Quarter was finally liberated on August 15 after holding out for 55 days and suffering more than 200 casualties. The relief of Peking broke the back of the Boxers and for all practical purposes the rebellion was over. American forces were to remain in northern China through the Fall of 1901 when peace negotiations were finally concluded.

The 11 U.S. ships involved in the China Relief Expedition include:

Brooklyn	Nashville	Wheeling
Buffalo	New Orleans	Yorktown
Iris	Newark	Zafiro
Monocacy	Solace	

The obverse of the original 400 China Relief Expedition Medals bears the date 1901. All subsequent medals, made from replacement dies, bear the date 1900. In 1913 a slight ribbon modification (from two thin black stripes to two thin blue edge stripes on the yellow ribbon) was introduced as reflected in the photos of this medal. Unprefixed rim numbers range from 1 to 400 on the 1901 strike and 401 to 1150 on the 1900 strike. There is also an early unnumbered strike of this medal. Marine Corps issues are numbered 1 to 600. All Marine Corps obverses bear the date 1900.

China Relief Expedition Medal (1900) Version I.

China Relief Expedition Medal (1900) Version II.

China Relief Expedition Medal reverse.

Navy Lieutenant John Cloy who won Medals of Honor in China (1900) and Vera Cruz, Mexico (1914) wears (left to right) the Navy Cross, Spanish Campaign Medal, Philippine Campaign Medal, China Relief Expedition Medal, Mexican Service Medal and WW I Victory Medal, circa 1920. Naval Historical Center.

Cuban Pacification Medal

Issued on August 13, 1909, this medal was originally authorized for all Navy personnel who served ashore in Cuba between October 6, 1906 and April 1, 1909. Following the Spanish American War, a government (independent of Spain) was formed. It was lead by President Thomas Estrada Palma. When he resigned in 1906, wide spread insurrection swept the country. With U.S. military support, a new government was formed under General Jose Miguel in 1908. While all land based forces were withdrawn by April 1, 1908, the Navy remained in the contiguous waters for an additional year to help ensure the stability of the new regime.

Eligibility was later extended to personnel serving aboard the following ships between September 12, 1906 and April 1, 1909:

Alabama	Dubuque	Minneapolis
Brooklyn	Eagle	Newark
Celtic	Illinois	New Jersey
Cleveland	Indiana	Paducah
Columbia	Iowa	Tacoma
Denver	Kentucky	Texas
Des Moines	Louisiana	Virginia
Dixie	Marietta	Prairie

This medal was issued with unprefixed rim numbers ranging from 1 to 2100. The Marine Corps version bears unprefixed numbers ranging from 1 to 1550.

Navy Cuban Pacification Medal.

Navy Cuban Pacification Medal reverse.

Nicaraguan Campaign Medal (1912)

During the decade of 1910, the government of President Jose Zelaya in Nicaragua proved to be most bothersome to U.S. financial interests. Zelaya spared nothing when it came to antagonizing Washington. His military adventurism in the region was a clear threat to stability in Central America. When the city of Bluefields revolted against him, he executed two American mercenaries fighting with the Bluefields' army. President William Howard Taft ordered the Pacific fleet to land a company of Marines under the command of the famous Major Smedley Butler. Butler succeeded in preventing Zelaya's forces from crushing the Bluefields' army. Zelaya was eventually forced into exile and a new government headed by Adolofo Diaz was installed.

In July of 1912, Zelaya overthrew Diaz's government. Rear Admiral H. H. Southerland was dispatched at the head of an eight ship squadron to restore order. Navy Lieutenant William D. Leahy and ten sailors were the first to go ashore at Corinto. Major Smedley

Continued on page 15

Butler returned with his Marines two days later on August 14, 1912. With additional reinforcements, Butler was ordered by Taft via Southerland to crush Zalaya's rebels. Supported by Nicaraguan forces loyal to Diaz, U.S. military personnel completely routed the rebel forces within two months and restored the Diaz government to power.

The first Nicaraguan Campaign Medal was issued to those personnel who served ashore or on the following eight ships between July 29 and November 14, 1912:

Annapolis	Denver
California	Glacier
Cleveland	Maryland
Colorado	Tacoma

The medal was issued with unprefixed rim numbers ranging from 1 to 1500. The unprefixed Marine Corps medals range from 1 to 1100.

First Nicaraguan Campaign (1912) Medal.

First Nicaraguan Campaign Medal reverse.

Mexican Service Medal

Mexico had been rocked by continual political unrest since 1910. This unrest led to serious tensions with the U.S. In a political coup, General Victoriano Huerta seized power. Relationships with President Woodrow Wilson's government quickly took a turn for the worse. In October of 1913, U.S. Admiral Frank Friday Fletcher was ordered to establish a naval presence from Vera Cruz to Tampico. Admiral Bradley Fiske subsequently ordered four battleships (Florida, Utah, Connecticut and Minnesota) to reinforce Admiral Fletcher. When eight members of an American shore party were arrested and jailed in Tampico, America immediately demanded their release and an apology. General Huerta released the men, but refused to provide an acceptable apology. When it was learned that the German ship Ypiranga was transporting war materials to Mexico, Wilson ordered the Marines to occupy Vera Cruz. On April 21, 1914, Marine Colonel Wendell Neville led a shore party of 500 Marines from the transport ship Prairie and Ensign George M. Lowry lead a shore party of 285 sailors from the Flagship Florida into Vera Cruz. These forces were supplemented by an additional battalion from the Utah under command of Ensign Paul Foster. They quickly overwhelmed any local resistance in bloody street fighting. All Navy personnel returned to their respective ships on April 27, turning over the continued occupation of Vera Cruz to U.S. Army forces. The Army occupation force did not withdraw

Continued on page 16

from Vera Cruz until November 23, 1914. Huerta's presidency was over and soon afterwards Venustiano Carranza seized power in Mexico.

Carranza however failed to curb the activities of the bandit Pancho Villa, who raided two New Mexican towns (Columbus and Ft. Furlong) in 1916. In response, President Wilson ordered General John "Black Jack" Pershing to invade Mexico with several regiments in retaliation. Numerous naval ships were stationed in the contiguous waters during this period to lend support to General Pershing's troops, if needed.

On February 11, 1918 the Mexican Service Medal was authorized for Navy personnel who served ashore at Vera Cruz in support of the military operations between April 21 and April 23, 1914. Eligibility was subsequently expanded to include personnel serving aboard 121 ships during the periods ranging from April 21 to November 26, 1914 and March 14, 1916 to February 7, 1917. There were also Navy personnel who served during periods of hostility outside of these dates, who also qualified for this medal.

It is one of only four Navy campaign medals awarded with prefixed ("M. No.") rim numbers (the Philippine Campaign, the Second Nicaraguan Campaign and the Yangtze Service Medals were the others). Unprefixed numbers ranged from 1 to 15499 and featured the curved "For Service" on the reverse. "M. No." prefixed rim numbers as produced by the U.S. Mint ranged from 15500 to 16674 and featured a straight "For Service" on the reverse. The Marine Corps version is unprefixed with a range of 1 to 2400. All Marine Corps issues use a curved "For Service."

The ships involved in the Mexican Campaign included:

Albany	Fanning	Morro Castle	Sacramento
Ammem	Florida	Nanshan	Salem
Annapolis	Flusser	Nashville	San Diego
Arethusa	Georgia	Nebraska	San Francisco
Arkansas	Glacier	Neptune	Saturn
Balch	Hancock	Nereus	Solace
Beale	Henley	Nero	Sonoma
Birmingham	Hopkins	New Hampshire	South Carolina
Brutus	Hull	New Jersey	South Dakota
Buffalo	Illinois	New Orleans	Sterrett
Burrows	Iris	New York	Stewart
California	Jarvis	North Dakota	Tacoma
Cassin	Jenkins	Ontario	Terry
Chatanoga	Jouett	Orion	Texas
Celtic	Jupiter	Ozark	Trippe
Chester	Kansas	Paducah	Truxton
Cheyenne	Kentucky	Patapsco	Utah
Cleveland	Lamson	Patterson	Vermont
Colorado	Lawrence	Patuxent	Vestal
Connecticut	Lebanon	Paulding	Vicksburg
Culgoa	Louisiana	Paul Jones	Virginia
Cummings	Machias	Perry	Vulcan
Cyclops	Marietta	Petrel	Walke
Delaware	Maryland	Pittsburgh	Warrington
Denver	Michigan	Prairie	Washington
Des Moines	Milwaukie	Preble	West Virginia
Dixie	Minnesota	Proteus	Wheeling
Dolphin	Mississippi	Raleigh	Whipple
Drayton	Monaghan	Reid	Wyoming
Eagle	Montana	Rhode Island	Yankton
			Yorktown

Navy Mexican Service Medal.

Navy Mexican Service Medal reverse.

Commander T. S. Wilkinson winner of the Medal of Honor at Vera Cruz, Mexico (1914) also wears (left to right) the Mexican Service Medal and WW I Victory Medal, circa 1935. Naval Historical Center.

Haitian Campaign Medal (1915)

Following a trouble free presidential election in Haiti in 1908, seven different individuals held the presidency over the next seven years. In 1913, General Guillaume Sam assumed power in a bloody coup. On no less than 16 occasions between late 1914 and early 1915 Marines went ashore to establish order. In July 1915, Ronsalvo Bobo led an uprising against General Sam. When Bobo's forces succeeded in overthrowing the General, President Wilson decided to intervene forcefully.

In the face of ongoing political unrest, President Wilson ordered Admiral William B. Caperton to lead three companies of sailors and two companies of Marines ashore at Port-au-Prince, Haiti on July 28, 1915 and restore order. The Marines made quick work of Bobo's rag-tag army. Haiti was destined to be governed by the U.S. Marines (who also formed and staffed a national police force) for most of the next two decades. The occupation would last until August, 1934.

The Haitian Campaign Medal was established on June 22, 1917. It bears the date 1915 on the obverse of the planchet. It was authorized for Navy personnel serving ashore between July 9 and December 6, 1915 or on board the following 14 ships during the same period:

Castine	Osceola
Celtic	Patuxent
Connecticut	Prairie
Culgoa	Sacramento
Eagle	Solace
Marietta	Tennessee
Nashville	Washington

Rim numbers for the Navy version ranged from 1 to 4000 and 4900 to 5300. There was no official Marine Corps version of this medal. Eligible Marines received medals with the Navy reverse in the 3000 range. Unnumbered medals with the distinctive Marine Corps reverse exist. These may be copies made for specific individuals or collectors or medals produced at some later date for bona fide Marine Corps recipients.

For recipients who later became eligible for the second Haitian Campaign Medal, a distinctive clasp with the dates "1919-1920" was authorized for wear on the suspension ribbon of the 1915 version of the medal.

First Haitian Campaign Medal (1915).

First Haitian Campaign Medal reverse.

First Haitian Campaign Medal with "1919-1920" clasp.

2

THE WORLD WAR I ERA

Dominican Campaign

The troubles in Haiti had spilled over into the Dominican Republic. With order restored in Haiti, Admiral Caperton was instructed to turn his attention there. The regime of Juan Jimenez was deeply in debt and troubled by the rebel forces of Desiderio Arias. President Wilson ordered Admiral Caperton to restore order in 1916. When negotiations broke down, Admiral Caperton sent the Marines in to seize control of the country. Led by double Medal of Honor recipient, Major Smedley Butler, the Marines quickly defeated the Dominican army. Caperton's successor, Rear Admiral Henry S. Knapp, then imposed martial law over the country on November 26, 1916 until a lawful government could be established.

Service in the Dominican Republic and on board the following 26 ships from May 16 through December 4, 1916 qualified for award of this medal:

Celtic	Hector	Prairie
Castine	Kentucky	Preston

Culgoa	Lamson	Reid
Dixie	Memphis	Sacramento
Dolphin	Machias	Salem
Eagle	Neptune	Solace
Flusser	Olympia	Sterrett
Hancock	Panther	Terry
	Potomac	Walke

The medal was authorized on December 29, 1921. The reverse design of both the Navy and Marine Corps version was the first to use a straight "For Service" exclusively. This reverse design would be used on all future Navy and Marine Corps campaign medals (except for the second Haitian Campaign medal) through the World War II era Navy Occupation Service Medal. Unprefixed rim numbers ranged from 1 to 3800 for the Navy issue. The Marine Corps issue ranged from 1 to 2800. Subsequent issues of this medal were unnumbered.

Dominican Campaign Medal.

Dominican Campaign Medal reverse. Note the initial usage of the straight "For Service" on the reverse.

Admiral William D. Leahy, who as a young Lieutenant figured prominently in the First Nicaraguan Campaign, wears (left to right) Navy Cross, Version II of the Sampson Commemorative Medal with USS Oregon brooch and four battle clasps, Spanish Campaign Medal, Philippine Campaign Medal, Mexican Service Medal, Nicaraguan Campaign Medal, Dominican Campaign Medal and the WW I Victory Medal, circa 1936. Naval Historical Center.

World War I Victory Medal

The assassination of Austria's Archduke Franz Ferdinand launched the first major war of the 20th century. Following the assassination, Austria-Hungry declared war on Serbia, which was quickly joined by its ally Russia. Soon thereafter Germany declared war on Russia and France and invaded Belgium, drawing Great Britain into the war against Germany. The war raged on in Europe with the U.S. avoiding being drawn into the conflict. Even the infamous sinking of the passenger liner Lusitania, along the Irish coast on May 7, 1915, with more than 120 American citizens on board, by the German U-Boat 20 did not manage to draw America into the war. In 1917 however, the Germans declared unrestricted submarine warfare against all Atlantic shipping. In March, German U-Boats sank five unarmed U.S. merchant vessels. President Woodrow Wilson could no longer maintain neutrality and on April 2, 1917 requested a declaration of war. Congress granted his request on April 6. After three years of deadly fighting with no strategic results, the U.S. entered the fray. U.S. armed forces grew to more than 4 million personnel. Approximately half saw combat action in Europe.

General John Pershing served as commander of the American Expeditionary Force (AEF). The first significant engagements involving American troops occurred in the summer of 1918. Continued successes into the early fall of 1918 turned the balance and in November, Germany surrendered. Wilson, hoping for an enlightened peace treaty, was dismayed by the petty bickering of the allied nations over its terms. The U.S. would enter into a long period of isolationism in disgust over the outcome. The U.S. refused to sign the Versailles Peace Treaty or to join the League of Nations, a forerunner of today's United Nations.

Due to the massive number of personnel who served in World War I, it was decided that a uniform medal would be authorized for all branches of the military service. Navy and Marine Corps personnel were authorized 19 distinctive service clasps as follows:

Armed Guard	Naval Battery
Asiatic	Overseas
Atlantic Fleet	Patrol
Aviation	Salvage
Destroyer	Subchaser
Escort	Submarine
Grand Fleet	Transport
Mine Laying	West Indies
Mine Sweeping	White Sea
Mobile Base	

Only one such clasp could be displayed on the suspension ribbon even though the individual may have qualified for multiple clasps. Among the more valuable clasps are: (1) Naval Battery; (2) Salvage; (3) White Sea; (4) Asiatic; (5) Aviation; and (6) Mobile Base.

In addition, Navy and Marine Corps personnel serving with the AEF in France, England, Italy or Russia were authorized to receive the following 11 Army distinctive battle and service clasps:

Aisne-Marne	Meuse-Argonne
Asine	Russia*
Defensive Sector	Siberia
England*	St. Mihel
France*	Ypres-Lys
Italy*	

The * denotes a "service" clasp as opposed to a "battle" clasp. Only one service clasp could be worn on the medal by any member of the armed forces. Multiple "battle" clasps, however could be worn on the suspension ribbon.

Members of the AEF who served in France were also authorized to wear a distinctive bronze Maltese cross on the suspension ribbon of the Victory medal. The Maltese cross could only be worn with a service clasp and not a battle clasp. It is usually observed being worn in combination with the Army's "France" service clasp.

Approximately 2,500,000 of these medals were originally issued to all services. The designer's name, Fraser, is stamped on the obverse of the medal. Due to the shear size of the issue the medals were unnumbered. An interesting fact is that a total of 14 nations issued WW I Victory medals. All used the same rainbow styled ribbon with their own distinctive planchet. In addition to the United States, the 13 nations who issued WW I Victory medals and the approximate number of medals issued were:

Country	Number Issued
Belgium	300,000
Brazil	2,500
Cuba	7,000
Czechoslovakia	90,000
France	2,000,000
Great Britain	6,000,000
Greece	200,000
Italy	Unknown
Japan	700,000
Portugal	100,000
Rumania	300,000
Siam (Thailand)	1,500
Union of South Africa	75,000

World War I Victory Medal.

World War I Victory Medal reverse.

Rear Admiral Cole receiving the Navy Cross from Rear Admiral William T. Tarrant. Note Admiral Cole's display (left to right) of the Sampson Medal with two engagement bars, Spanish Campaign Medal, Philippine Campaign Medal and World War I Victory Medal with clasp, circa 1940. Emering Collection.

Army of Occupation Medal

This medal (basically an Army medal) established on November 21, 1941 was issued to those Naval and Marine Corps personnel serving with the occupation forces in Germany or Austria-Hungary, following Germany's defeat in World War I, between November 12, 1918 and July 11, 1923. The obverse features the profile of General John J. Pershing, commander of the AEF in Europe. The medal underwent a minor ribbon change prior to its original issue. It is believed that none of the medals were ever officially issued with the original ribbon. The original ribbon is available, however and examples of the original configuration do exist.

Army of Occupation of Germany Medal awarded to sailors and Marines serving at land bases in Germany or Austria-Hungary following World War I.

Army of Occupation of Germany Medal reverse.

Haitian Campaign Medal (1919-1920)

Following the original occupation of Haiti in 1915, the native Cacos, under the leadership of Charlemagne Paeralte, started a wave of violence in 1918 directed at the overthrow of the U.S. supported government. As a result, additional Navy and Marine personnel were rushed to the island in a protracted effort to restore law and order. Charlemagne was eventually assassinated in a cleverly staged ambush near Fort Riviere in October 1919 by Marine Sergeant (later Brigadier General) Herman H. Hanneken. Hanneken was serving as a Captain in the Haitian Gendarmerie (national police) at the time. With the loss of Charlemagne, and no one of equal stature to replace him, the rebellion was effectively broken. The Marine Corps occupation would continue until 1934.

The second Haitian Campaign Medal was awarded to Navy and Marine Corps personnel from the following 37 ships, which served in Haitian waters between April 1, 1919 and June 15, 1920:

Beaufort	Rowan
Delaware	Sandpiper
Dolphin	Shubrick
Gulfport	Subchaser 135
Hancock	Subchaser 136
Henderson	Subchaser 165
Kittery	Subchaser 180
Kwasind	Subchaser 210
Lake Bridge	Subchaser 211
Lake Worth	Subchaser 212
Long Beach	Subchaser 213
May	Subchaser 214
Mercy	Subchaser 223
Mohave	Subchaser 251
Osceola	Subchaser 253
Pensacola	Subchaser 443
Peoria	Subchaser 444
Potomac	Winslow
Prometheus	

The medal is identical in all respects to the first Haitian Campaign Medal, except that the obverse of the planchet bears the dates 1919-1920 instead of 1915. Navy and Marine Corps personnel, who received the first Haitian Campaign Medal for service during 1915, were awarded a distinctive clasp for the first medal with the dates "1919-1920" to be worn on the suspension ribbon of the original 1915 medal. All Navy issues of the second medal were unnumbered and utilized a curved "For Service." Marine Corps issues used unprefixed numbers ranging from 1 to 3250. Unofficial strikes of this medal bearing the date "1919" on the planchet also exist.

Second Haitian Campaign Medal (1919-1920) obverse.

Second Haitian Campaign Medal reverse.

Marine Brigadier General Herman Hanneken, who as a Sergeant (pictured here) assassinated the rebel leader, Charlemagne Paeralte, near Fort Riviere, Haiti bringing to end the unrest, which led to the Second Haitian Campaign. National Archives.

Marine Corps Expeditionary

Heretofore, the Marine Corps had "dove tailed" off of the Navy campaign medal system, using the Navy's obverse with a distinctive Marine Corps reverse. This medal, established in July, 1921, represented an interesting divergence in the established system. This was the first medal to use a distinctive Marine Corps obverse, but it still utilizes the standard Marine Corps reverse with a straight "For Service." The suspension ribbon reflects the Marine Corps colors of red and gold. Its intent was, and continues to be, recognition of Marine Corps landings in hostile territory as part of an operation against an armed enemy, which do not qualify for any other campaign medal. The award of this medal initially recognized many past operations, dating as far back as 1874, for which no other campaign medal had been awarded. In fact 45 of the 66 operations, which have qualified for award of this medal, occurred prior to authorization of this medal in 1921. The qualifying operations and the initial year of each are listed here as a matter of historical interest.

Operation	Year		
Honolulu	1874	Seoul	1896
Alexandria, Egypt	1882	Tientsin, China	1894
Panama	1885	Chefoo, China	1895
Seoul, Korea	1888	Colombia	1895
Samoa	1888	Corinto, Nicaragua	1898
Honolulu	1889	San Juan, Nicaragua	1898
Buenos Aires	1890	Peking, China	1898
Haiti	1891	Bluefields, Nicaragua	1899
Chile	1891	Samoa	1899
Honolulu	1893	Panama	1901
Bluefields, Nicaragua	1894	Colombia	1902

Isthmus of Panama	1902	Haiti	1920
Panama	1902	Turkey	1921
Honduras	1903	Turkey	1922
Dominican Republic	1903	Honduras	1924
Syria	1903	Peking, China	1924
Panama	1903	Shanghai, China	1925
Abyssinia	1903	Wuchow, China	1926
Dominican Republic	1904	Bluefields, Nicaragua	1926
Seoul, Korea	1904	Canton, China	1927
St. Petersburg, Russia	1905	China	1928
Honduras	1907	Haiti	1931
Bluefields, Nicaragua	1909	Shanghai, China	1934
Corinto, Nicaragua	1910	Wake Island	1941
Peking, China	1911	Cuba	1961
Shanghai, China	1911	Thailand	1962
Guantanamo Bay, Cuba	1912	Iran/Yemen	1978
Managua, Nicaragua	1912	Panama	1980
Haiti	1914	Lebanon	1982
Dominican Republic	1914	Libya	1986
Haiti	1915	Persian Gulf	1987
Dominican Republic	1916	Liberia	1990
Haiti	1918	Rwanda	1994

The initial issues of this medal produced by the U.S. Mint bore the "M. No." prefix ranging from 1 to 10000. Later strikes of this medal are unnumbered. The 1941 Wake Island operation qualified for display of the distinctive bronze "Wake Island" clasp on the suspension ribbon and a 1/4th inch silver "W" on the corresponding service ribbon. Additional awards of this medal are noted by wear of a 3/16th inch bronze star on the suspension ribbon. A 3/16th inch silver star is used to denote five awards of this medal. This is the last medal with the distinctive Marine Corps reverse, which continues to be available for issue.

Marine Corps Expeditionary Medal obverse.   *Marine Corps Expeditionary Medal reverse.*

Second Nicaraguan Campaign

In 1925 the U.S. began withdrawing its troops from Nicaragua, the scene of an earlier banana war during the 1910s. Seizing the opportunity, General Jose Moncada revolted against the incumbent government. Again, finding it necessary to protect American (financial) interests during a period of political unrest, the Navy and Marines were sent to Nicaragua for a second time in January 1927. Secretary of War Henry Stimson followed close on their heels. He was able to negotiate a peaceful settlement between the government and General Moncada. One of the younger rebel leaders, Augusto Sandino, refused to agree and along with his followers, known as Sandinistas, launched what would be the last of the U.S.'s banana wars on July 16, 1927.

A bloody five year war would be fought mostly in the mountainous region near the Honduran border between U.S. Marines and the Sandinistas. By 1933, with the conflict winding down and a new President, Juan Sacasa, in office the U.S. pulled out of Nicaragua. On February 2, 1933, the Sandinistas agreed to a complex peace agreement. When difficulties arose in early 1934, Sandino was called to the Presidential Palace in Managua for discussions. As the unsuccessful negotiations concluded, he and his top aids were arrested and assassinated by General Anastasio Somoza. Somoza and his family would rule Nicaragua from 1937 through 1979, when the government would be overthrown by a new generation of Sandinistas.

A second campaign medal was authorized for service between August 27, 1926 and January 2, 1933, when the last contingent left Nicaragua. The ribbon is very distinctive from that of the original campaign medal. All initial Navy issues bore an "M. No." prefix with the numbers ranging from 1 to 10000. Marine Corps issues utilized "M. No." numbers ranging from 1 to 6000.

Personnel on board the following ships during the aforementioned period also qualified for the medal:

Ashville	Goff	Milwaukie	Shirk
Bainbridge	Hatfield	Mullany	Sloat
Barker	Henderson	Osborne	Smith Thompson
Barry	Humphreys	Overton	Sturtevant
Borie	Kane	Paulding	Tracy
Brooks	Kidder	Philip	Trenton
Cincinnati	King	Preston	Tulsa
Cleveland	La Vallette	Quail	Whipple
Coghlan	Lawrence	Raleigh	Wickes
Denver	Litchfield	Reid	Williamson
Detroit	Marblehead	Reuben James	Wood
Edwards	Marcus	Robert Smith	Yarborough
Flusser	McFarland	Rochester	
Galveston	Melvin	Sacramento	
Gilmore	Memphis	Selfridge	

Second Nicaraguan Campaign Medal obverse.

Second Nicaraguan Campaign Medal reverse.

Navy Lieutenant (jg) Cameron Winslow, Jr. wearing the Navy Cross and Second Nicaraguan Campaign Medal, circa 1936. Emering Collection.

Yangtze Service Medal

Following the Boxer Rebellion at the turn of the century, the U.S., under the favorable terms of the ensuing peace treaty, established a substantial commercial presence in and around Shanghai. Numerous other nations followed suit. The treaty also provided for a U.S. Naval gunboat presence on the local rivers.

A wave of Chinese nationalism beginning in 1924 sought to shed light on the unfairness of the treaty and the need for stronger nationalism. This unrest exploded into violent student street riots in 1925 and threatened U.S. (financial and commercial) interests in China. This violence was followed by a number of small scale punitive landings by U.S. forces on Chinese territory. In 1926, Chiang Chieh-shih, known as Chiang Kai-shek in the West, and his nationalist forces laid siege to Hankow, which was defended by the Chinese Warlord Wu P'ei-fu. Alarmed by these events, Rear Admiral Henry Hough stationed two gunboats and the Destroyers Pope and Stewart in the harbor at Hankow. Wu's surrender led to a subsequent series of confrontations between Chiang's Nationalist forces and the gunboat patrols of the British and U.S. Navies. When the confrontations finally abated, the Nationalist forces marched on Shanghai.

In the spring of 1927, Chiang's forces attacked the city and riots erupted inside the International Settlement at Shanghai. A multinational force, comprised in part of 1,700 Navy and Marines, under the command of now Brigadier General Smedley Butler, were sent to protect the City of Shanghai and several other areas with material U.S. business interests.

In early 1932, tensions flared between the Japanese and Chiang's Nationalists following Japan's conquest of Manchuria. This erupted into relatively large scale fighting in and around Shanghai. The Japanese even used dive bombers against the Chinese district of Chapei. Again, the safety of the International Settlement was imperiled. President Hoover dispatched additional ships to Shanghai. They were joined by the British fleet and together they pressured the Japanese to agree to a temporary ceasefire. For the time

being the "Open Door" policy with China would remain in effect during a very fragile peace.

Established on April 28, 1930, the Yangtze Service Medal was authorized for Navy and Marine Corps personnel serving in the Yangtze River Valley between September 3, 1926 and October 21, 1927 and/or March 1, 1930 and December 31, 1932. Soldiers from the Army's 31st Infantry also qualified for this award and were presented the U.S. Marine variant. In addition, landing parties from the Chaumont and the crews of the following 69 ships qualified for this medal:

Alva	Henderson	Peary	Sub-36
Asheville	Heron	Pecos	Sub-37
Avocet	Houston	Penguin	Sub-38
Barker	Hulbert	Pigeon	Sub-39
Beaver	Isabel	Pillsbury	Sub-40
Bittern	Jason	Pittsburgh	Sub-41
Black Hawk	Jones	Pope	Sacramento
Borie	Luson	Preble	Sicard
Bulmer	MacLeish	Preston	Simpson
Canopus	Marblehead	Pruit	Stewart
Cincinnati	McCormick	Richmond	Thompson
Edsell	Monocacy	Rizal	Tracy
Edwards	Noa	Rochester	Truxtun
Elcano	Oahu	Sub 30	Tulsa
Finch	Palos	Sub-31	Tutuila
Ford	Panay	Sub-34	Villalobos
Guam	Parrett	Sub-35	Whipple
Hart			

The Yangtze Service Medal was produced by the U.S. Mint and issued with prefixed "M.No." 1 to 13500 for Navy personnel and 1 to 6500 for Marine Corps personnel and utilized a straight "For Service" on the reverse.

Yangtze Service Medal obverse.

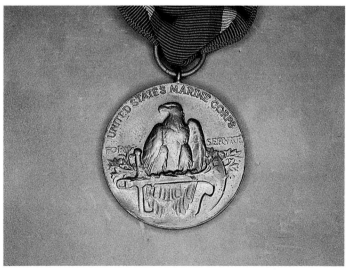

Yangtze Service Medal reverse.

Major General Smedley D. Butler, double Medal of Honor winner (Mexico 1914 and Haiti 1915), figured prominently in America's banana wars during the first half of the 20th century. General Butler served in Cuba (1906), Nicaragua (1912), Mexico (1914), Haiti (1915), Dominican Republic (1916) and in China (1927). National Archives.

3

THE WORLD WAR II ERA

Navy Expeditionary Medal

Instituted on August 5, 1936 this medal is issued for actual engagement of the enemy in combat situations ashore, or for specially meritorious operations for which no corresponding campaign medal exists. Like its Marine Corps counterpart, the vast majority of operations for which this medal was issued occurred prior to its creation. Its main function was to make up for past oversights. The suspension ribbon reflects the Navy colors of blue and gold. For those Navy personnel who participated in the defense of Wake Island between December 7 and 22, 1941 authorization was provided for display of a bronze "Wake Island" clasp on the suspension ribbon and a 1/4th inch silver "W" on the corresponding service ribbon. This medal continues in use today under the aforementioned circumstances. Although it has been authorized as recently as the 1990s, the Armed Forces Expeditionary Medal serves virtually the same purpose and has been used when other branches of the service have been involved. This seems, for the most part, to be more the rule than the exception in the modern era. It is the last campaign medal with the distinctive Navy reverse (straight "For Service"), which continues to be available for issue.

The 59 engagements which qualified for award of the Navy Expeditionary Medal through World War II include:

China	1894, 1895, 1898, 1911, 1924, 1925, 1926, 1927, 1928 and 1937
Cuba	1912
Dominican Republic	1903, 1904, 1914, 1916 and 1918
Egypt	1882
Haiti	1891, 1914, 1915, 1918, 1920 and 1929
Hawaii	1874, 1889, and 1893
Honduras	1903, 1907 and 1924
Korea	1888, 1894, 1895 and 1904
Nicaragua	1894, 1896, 1898, 1899, 1909, 1910, 1912 1915, 1918 and 1926
Panama	1885, 1901, 1902 and 1903
Philippine Islands	1911
Russia	1905
Samoa	1888 and 1899
Siberia	1920
Syria	1903
Turkey	1921 and 1922
Wake Island	1941

Abyssinia	1902
Argentina	1890
Chile	1891

The eight engagements qualifying for award of the Navy Expeditionary Medal after World War II include:

Cuba	1961	Lebanon	1982
Iran	1978 and 1979	Liberia	1990
		Libya	1986
		Persian Gulf	1987
		Rwanda	1994
		Thailand	1962

Navy Expeditionary Medal with Wake Island clasp obverse.

Navy Expeditionary Medal reverse.

China Service Medal

Japan's expansionist programs in the Far East traces its roots back to the 1931 invasion of Manchuria. This served as a platform for a future incursion into China. By the late 30s, Japan had stepped up its aggression against Chiang Kai-shek's Nationalist government. During the summer of 1937, Admiral Kiyoshi Haseawa would launch air and naval attacks against Nanking. The Navy's presence consisted of approximately 17 ships and a few small gunboats. There were also about 2,300 soldiers and Marines ashore. Their purpose was to protect American diplomatic and financial interests.

In the fall of 1937, the Japanese Army occupied Nanking. There, the Japanese forces randomly killed the populace in an action that would become known as the "Rape of Nanking." It is estimated that more than 300,000 people were killed at Nanking. Meanwhile, Chiang had fallen back inland to Hankow. On December 12, 1937, with Hankow under siege, the Japanese Air Force bombed and sank the American Gunboat Panay, killing three and wounding 48, including five civilians.

When Hankow fell, Chiang fell back even further to Chungking. An unprecedented apology from the Japanese emperor helped diffuse American anger surrounding the Panay incident and our Navy continued to standby during this period of hostilities. Chiang managed to stalemate the Japanese at Chungking and they eventually turned their attention to consolidating their conquests in Manchuria and China. This stalemate played a key role in limiting overall Japanese efforts in Southeast Asia during World War II.

In 1939, Japan stunned the world by announcing a three party agreement with Germany and Italy. Little did the U.S. realize what was still yet to come in the Asiatic-Pacific region.

The China Service Medal was originally issued to Navy personnel serving ashore or on vessels in the contiguous waters during the period of Japanese aggression against Nationalist China from July 7, 1937 to September 7, 1939.

The China Service Medal was also awarded for post World War II service from September 2, 1945 to April 1, 1957. During this latter period, confusion and chaos reigned inside China. During the early 40s, Japan established a puppet government under Wang Ching-wei at Nanking. In late 1944, Japanese forces inflicted a major defeat on Chiang's Nationalist forces.

The eventual defeat of Japan in 1945 left China with no less than five separate armies in its territory. These included the Russians, U.S. Army and Marines, the defeated Japanese forces, the Chinese Nationalist Army and the Chinese Communist forces. The Russians had been lured into the Far Eastern conflict by promises of economic concessions offered to Stalin at the infamous Yalta Conference by President Roosevelt and Prime Minister Churchill.

In 1946, U.S. Marines saw heavy combat action against regional Communist forces at Anpin, Hopeh. Additional action would follow a few months later with a Communist attack on the Marine ammunition depot at Hsin Ho. The Marines had completely evacuated Hopeh by April 1947. All remaining Marines would leave China in May 1948. U.S. Navy forces off mainland China would experience a similar reduction to about a dozen cruisers and destroyers.

As 1948 progressed, MaoTse-tung's Red Army began making significant gains against Chiang's Nationalist forces. By early 1949, the Nationalist forces were seeking a negotiated settlement with Mao. By October 1, 1949, Mao proclaimed the establishment of the People's Republic of China. The "China Open Door" policy was dead and would remain so for the next three decades. By December 8, 1949, Chiang's Nationalist forces had been driven across the Straits of Formosa to the Island of Taiwan. Here Chiang established the Republic of China, quickly declaring martial law. This was done as much to protect the new government from Mao's forces as much as it was to protect it from the native Taiwanese, who culturally had been heavily influenced by the Japanese. Chiang would rule Taiwan as a virtual dictator until his death in 1978 when he would be succeeded by his son, Chiang Ching-kuo.

On June 25, 1950 the invasion of South Korea by North Korea insured support for the continued existence of the Taiwan government, by the U.S., which recognized Taiwan's strategic importance. Elements of the Seventh Fleet would remain on the scene to protect U.S. interests well into the mid-50s.

If an individual qualified during both periods of service, a 3/16th inch bronze star was authorized for display on the medal's suspension ribbon. The initial award of this medal (1937 to 1939) was issued unnumbered to Navy personnel and used non-prefixed rim numbers ranging from 1 to 4000 for eligible Marine Corps recipients. The second issue was awarded unnumbered to both Navy and Marine Corps recipients. A total of 45 ships qualified for award of the China Service Medal during the first period (1937-1939).

First Period

Alden	Guam	Pillsbury
Asheville	Henderson	Pope
Augusta	Heron	Ramapo
Barker	Isabel	Sub-36
Bittern	Jones	Sub-37
Blackhawk	Luson	Sub-38
Bridge	Marblehead	Sub-39
Bulmer	Mindanao	Sub-40
Canopus	Monocacy	Sub-41
Chaumont	Oahu	Sacramento
Edsell	Panay	Stewart
Edwards	Parrott	Trinity
Finch	Peary	Tulsa
Ford	Pecos	Tutuila
Gold Star	Pigeon	Whipple

Given the length of the second period (1945 - 1957) more than 1,000 ships qualified for award of the China Service Medal.

Second Period

Abbot	Dade	LCI 687	LST 762	Rupertus
Achelous	Davis	LCI 704	LST 772	Rush
Achernar	Davison	LCI 705	LST 774	Ryer
Acree	Deal	LCI 706	LST 799	Sabalo
Adair	Dehaven	LCI 762	LST 802	Sabine
Adonis	Deliver	LCI 764	LST 803	Safeguard
Adria	Delta	LCI 766	LST 804	St. Croix
Agawam	Derrick	LCI 767	LST 845	St. Paul
Agerholm	Dextrous	LCI 769	LST 846	Salamonie
Ahrens	Diachenko	LCI 770	LST 854	Salisbury Sound
Albuquerque	Diomedes	LCI 787	LST 885	Samar
Alderbaran	Dixie	LCI 801	LST 866	San Clemente
Alderamin	Doyle	LCI 802	LST 898	Sandoval
Algol	Duluth	LCI 803	LST 919	San Saba
Alhena	Duncan	LCI 804	LST 945	Santee
Almack	Dunlin	LCI 805	LST 982	Sarasota
Alshain	Dutton	LCI 806	LST 985	Sarita
Alstede	Duxbury Bay	LCI 965	LST 1027	Sarpendon
Aludra	Dyes	LCI 967	LST 1078	Satyr
Ammen	Eaton	LCI 969	LST 1089	Savage

Anderson	Effingham	LCI 970	LST 1102	SC 1060
Andromeda	Eldorado	LCI 981	LST 1130	SC 1349
Antares	Elkhorn	LCI 985	LST 1141	Scabardfish
Anzio	Embattle	LCI 989	LST 1159M	Schuyler
APC 18	Endicott	LCI 1054	Lubbock	Scribner
APL 10	Energy	LCI 1084	Lucidor	Sea Cat
APL 11	Entemedor	LCI 1090	Mack	Sea Devil
APL 29	Epperson	LCI 1092	Mackenzie	Sea Dog
APL 44	Erben	LCS 4	Maddox	Sea Fox
Aquarius	Essex	LCS 8	Magoffin	Secota
Arcturus	Estero	LCS 11	Mainstay	Seekonk
ARD 31	Estes	LCS 12	Makin Is.	Seiverling
Arequipa	Evans	LCS 13	Malabar	Seminole
Arikara	Everett	LCS 14	Manatee	Serrano
Arneb	Eversole	LCS 18	Manchester	Sevier
Arundel	Fall River	LCS 22	Manokin	Sgt. Muller
Ascella	Faribault	LCS 27	Mansfield	Shnagri La
Ashtabula	Fearless	LCS 28	Marias	Shellbark
Askari	Fechteler	LCS 29	Markab	Shelton
Astoria	Fieberling	LCS 30	Marsh	Shields
ATA 122	Finch	LCS 31	Marshall	Shoveler
ATA 181	Firedrake	LCS 32	Mason	Siboney

Atlanta	Fiske	LCS 33	Matanikau	Sierra
ATR 62	Fletcher	LCS 34	Mathews	Silverbell
ATR 72	Florikan	LCS 35	Mattaponi	Silverstein
ATR 87	Floyds Bay	LCS 36	Maumee	Sirona
ATR 181	Formoe	LCS 43	McCaffrey	Sitka
Aucilla	Ft Marion	LCS 44	McDermut	Skagit
Ault	Foss	LCS 45	McGinty	Small
Auman	Fox	LCS 46	McIntyre	Smith
Badoeng	Frament	LCS 47	McKean	Soley
Balduck	Gage	LCS 48	Mellena	Southerland
Baltimore	Gantner	LCS 50	Menard	Spangler
Bandera	Gardiners Bay	LCS 51	Menhaden	Spencer
Barbican	Gary	LCS 52	Menifee	Spinax
Barton	Gatling	LCS 53	Merapi	Sprig
Bashaw	Gavia	LCS 54	Merrick	Springfield
Bass	Gendreau	LCS 55	Merrimack	Sproston
Bausell	Gen. Anderson	LCS 56	Mervine	Starlight
Bayfield	Gen. Blatchford	LCS 57	Metivier	Starr
Beaver	Gen. Breckenridge	LCS 58	Midway	Stembel
Becuna	Gen. Butner	LCS 59	Mills	Stentor
Begor	Gen. Ettinge	LCS 60	Mindanao	Sterlet
Belle Grove	Gen. Mann	LCS 70	Minooka	Stickell
Bellerophon	Gen. Mitchell	LCS 71	Mispillion	Stoddard
Beltrami	Gen. Muir	LCS 72	Mission Purisma	Strong
Benevolence	Gen. Randall	LCS 73	Missouri	Suffolk
Benham	Gen. Sulton	LCS 74	Moale	Svisun
Benner	Geneva	LCS 75	Moctobi	Summer
Bennington	George	LCS 76	Molala	Surfbird
Bergall	Gilmer	LCS 77	Monrovia	Sussex
Bergen	Golden City	LCS 78	Monssen	Swenson
Berrien	Goodrich	LCS 96	Montague	Swift
Berry	Gosselin	LCS 97	Montauk	Symbol
Besugo	Graffias	LCS 100	Montrose	Tabberer
Bexar	Grainger	LCS 102	Moore	Tulladega
Black Hawk	Grapple	LCS 103	Motive	Taluga
Blenny	Greenlet	LCS 104	Mountrail	Tangier
Blue	Greenwich Bay	LCS 105	Mulberry	Tantalus
Bluegill	Gregory	LCS 106	Mullany	Toppahannock
Bluebac	Grundy	LCS 107	Munro	Tarawa
Bluebird	Guadalupe	LCS 108	Munsee	Tate
Boarfish	Gudgeon	LCS 126	Murrelet	Tatum
Bole	Gunason	LCT 403	Nabigwon	Taussig
Bolivar	Gurke	LCT 404	Nacheninga	Tawakoni
Bordelon	Hale	LCT 514	Naief	Tawasa
Bougainville	Hamner	LCT 515	Nahasho	Taylor
Boxer	Hampton	LCT 892	Nomakagon	Tazwell
Boyd	Hancock	LCT 1004	Nantahala	Tekesta
Boyle	Hanna	LCT 1171	Napa	Telfair
Bracken	Hanson	LCT 1284	Nashville	Thomas
Bradford	Hassayampa	LCT 1330	Natchaug	Thomason
Braine	Haverfield	LCT 1333	Navasota	Thompson
Bremerton	Hawkins	LCT 1335	Neches	Thuban
Brevard	Hector	LCT 1343	New Jersey	Tingey
Brisco	Helena	LCT 1364	Newport	Tinsman
Brister	Henderson	LCT 1389	Nichola	Tiru

Bristol	Henrico	LCT 1411	Nicollet	Titania
Bronx	Herndon	LCT 1412	Noble	Tlingit
Brookings	Hickox	LCU 877	Norris	Todd
Brown	Hidalgo	LCU 1236	Notable	Token
Brule	Higbee	LCU 1446	Oak Hill	Toledo
Brush	Hisada	Lenawee	Oakland	Tolland
Buck	Hollister	Leo	O'bannon	Tollberg
Buckley	Holton	Leon	Oberon	Tolovania
Bugara	Hooper Island	Lignite	O'Brien	Tolowa
Bull	Hopewell	Lipan	Obstructer	Tombigbee
Bumper	Hornet	Lofberg	Oconto	Tomich
Burleson	Hubbard	Long	Octans	Topeka
Cabezon	Hunt	Los Angeles	Okanogan	Torchwood
Cabildo	Huntington	LSM 3	Oneida	Tortuga
Cacapon	Hyades	LSM 14	Onslow<	.Toucan
Caelum	Hyde	LSM 35	Orca	Towaliga
Calamares	Hydrus	LSM 43	Oriskany	Towner
Caliente	Hyman	LSM 55	Orleck	Trathen
Calvert	Indian Island	LSM 62	Ortolan	Tucker
Cambria	Indra	LSM 69	Orvetta	Tutulia
Cape Esperance	Ingersoll	LSM 74	Osmus	Typhon
Cape Johnson	Ingham	LSM 76	Ostara	Uhlmann
Capricornus	Ingraham	LSM 112	Owen	Ulysses
Carbonero	Iolanda	LSM 118	Ozbourn	Union
Carib	Ibex	LSM 124	Palawan	Uvalde

Carmick	Isbell	LSM 155	Parks	Valentine
Carp	Isherwood	LSM 157	Pasadena	Valley Forge
Carpenter<	Jaccard	LSM 173	Passumpsic	Vammen
Carroll	Jarvis	LSM 174	PC 491	Venango
Carter Hall	Jason	LSM 208	PC 498	Vesole
Carteret	Jefferson	LSM 218	PC 575	Vestal
Casa Grande	Jenkins	LSM 228	PC 593	Volador
Caswell	Jicarilla	LSM 225	PC 594	Volans
Catamount	Juneau	LSM 248	PC 802	Wabash
Catclaw	Jupiter	LSM 249	PC 804	Wadleigh
Catfish	Kadashan Bay	LSM 250	PC 807	Wahoo
Catoctin	Karin	LSM 251	PC 1079	Wakefield
Cavalier	Karnes	LSM 256	PC 1134	Walke
Centaurus	Kaskaskia	LSM 274	PC 1142	Walker
Cepheus	Kearsage	LSM 279	PC 1546	Walton
Chandler	Keith	LSM 282	PCE 855	Waneta
Chara	Kenmore	LSM 293	PCS 1455	Wantuck
Charleston	Kennebec	LSM 314	Peacock	War Hawk
Charr	Kennedy	LSM 335	Perkins	Warren
Charrette	Kephart	LSM 336	Philip	Warrick
Chase	Keppler	LSM 340	Philippine Sea	Washburn
Chauncey	Kerstin	LSM 341	Phoebe	Wasp
Chemung	Kidd	LSM 348	Pickaway	Watts
Chevalier	Kimberly	LSM 350	Pickerel	Waukesha
Chicago	Kinzer	LSM 351	Pictor	Warbill
Chickasaw	Kishwaukee	LSM 364	Piedmont	Waxwing
Chikaskia	Kite	LSM 373	Pine Is.	Wedderburn
Chilton	Kittson	LSM 375	Pirate	Wharton
Chimaera	Knapp	LSM 378	Pitt	Whetstone
Chimariko	Knox	LSM 393	Pittsburgh	Whidbey
Chincoteague	Knudson	LSM 431	Pivot	Whipstock
Chipola	Koiner	LSM 433	Platte	Whiteside
Chiquito	Kretchmer	LSM 444	Pt. Cruz	Whitting
Cholocco	Kukui	LSM 450	Polana	Willett
Chopper	Kula Gulf	LSM 457	Polaris	Wilson
Chowanoc	Kyes	LSM 460	Pollux	Wiltsie
Chubb	Lacerta	LSM 462	Pomfret	Windham Bay
Cimarron	Larson	LSM 475	Pondera	Winged Arrow
Cinnamon	Latona	LSM 482	Porter	Winston
Clarendon	Lavaca	LSM 484	Porterfield	Widgeopn
Clarion	Lawe	LSM 488	Potter	Wood
Claxton	Laws	LSM 489	Powell	Woodson
Clearfield	LC 362	LSMR 409	Prairie	Worcester
Cliffrose	LC 370	LSMR 536	Pres. Adams	Xanthus
Clinton	LC 485	LST 26	Pres. Jackson	Yancey
Clymer	LC 575	LST 42	Prichett	Yarnall
Cockrell	LC 988	LST 45	Princeton	Yavapai
Cogswell	LCI 21	LST 49	Prometheus	YFN 691
Chocton	LCI 22	LST 75	Ptarmigan	YFN 736
Colahan	LCI 23	LST 125	Puget Sound	YFN 746
Colbert	LCI 31	LST 172	Quapaw	YFN 755
Cole	LCI 195	LST 219	Queenfish	YMS 330

Collett	LCI 196	LST 226	Quick	YMS 336
Colonial	LCI 225	LST 229	Raby	YMS 342
Columbus	LCI 233	LST 267	Radford	YMS 346
Comet	LCI 234	LST 308	Ramsden	YMS 370
Comstock	LCI 338	LST 372	Rankin	YMS 386
Conner	LCI 342	LST 452	Raton	YMS 387
Connolly	LCI 446	LST 466	Razorback	YMS 396
Conserver	LCI 448	LST 469	Redstart	YMS 398
Constant	LCI 607	LST 485	Regulus	YMS 414
Conway	LCI 608	LST 494	Rehoboth	YMS 442
Cony	LCI 609	LST 505	Remey	YMS 468
Corduba	LCI 610	LST 516	Rendova	YOG 77
Cormorant	LCI 611	LST 539	Renshaw	Yokes
Coronis	LCI 612	LST 565	Renville	Yorktown
Corson	LCI 613	LST 602	Repose	Young
Cortland	LCI 616	LST 611	Reynolds	YP 627
Coucal	LCI 630	LST 616	Richey	YP 631
Cowell	LCI 631	LST 618	Ringness	YP 646
Crag	LCI 632	LST 620	Rio Grande	YTB 518
Craig	LCI 632	LST 629	Rochester	YTB 521
Cree	LCI 638	LST 640	Rock	YW 90
Cronin	LCI 639	LST 641	Rockbridge	YW 105
Cullman	LCI 647	LST 665	Rocky Mountain	Zelima
Cummings	LCI 648	LST 697	Rogers	
Cunningham	LCI 649	LST 706	Roi	
Current	LCI 658	LST 715	Romulus	
Currier	LCI 683	LST 735	Ronquil	
Currituck	LCI 685	LST 744	Roosevelt	
Cushing	LCI 686	LST 758	Rowan	

China Service Medal obverse.

China Service Medal reverse.

Navy Chief Machinist Mate William Badders who won a peace time Medal of Honor at Portsmouth, New Hampshire during the rescue and salvage of the sunken submarine, USS Squalus, also wears (left to right) the Navy Cross, the WW I Victory Medal with the Atlantic Fleet Service Clasp, the Yangtze Service Medal and the Navy Good Conduct Medal, circa 1940. Naval Historical Center.

American Defense Service Medal

Authorized on June 28, 1941, this medal recognizes service in the armed forces during the emergency period, occasioned by the breakout of hostilities in Europe, proclaimed by President Franklin D. Roosevelt from September 8, 1939 to December 7, 1941. The Navy required a minimum of 10 days active service during the aforementioned period in order to qualify for the medal. In addition, the Navy authorized the award of two distinctive clasps: (1) "Base" for service at bases outside the contiguous United States; and (2) "Fleet" for sea service or as a member of an aircraft squadron. In addition,

a bronze, block-style 1/4th inch block style "A" was authorized for wear on the service ribbon for personnel serving on ships operating in the North Atlantic in close proximity to Axis powers, i.e. in danger of U-Boat attack, between June 22 and December 7, 1941. This medal is unnumbered and uses a reverse common to all branches of the service. The reverse reads: "For Service During the Limited Emergency Proclaimed By The President On September 8, 1939 Or During The Unlimited Emergency Declared By The President On May 27, 1941."

American Defense Service Medal obverse.

American Defense Service Medal reverse.

American Defense Service Medal obverse with two authorized Navy clasps: "Base" and "Fleet."

American Campaign Medal

On November 6, 1942 Executive Order 9265 divided the world into three distinct Theaters of Operations for World War II. The American Campaign Medal was awarded for service during the period December 7, 1941 and March 2, 1946 in the American Theater of Operations. The American Theater of Operations included Canada, the contiguous United States and South America. Four separate campaigns qualified for a 3/16th inch bronze star:

Antisubmarine
Armed Guard
 Escort
Special Operations

American Campaign Medal obverse.

American Campaign Medal reverse.

U.S. warships destroyed at Pearl Harbor by Japanese fighter bombers. December 7, 1941. This act of infamy made America's entry into World War II Inevitable. National Archives

Asiatic-Pacific Campaign Medal

The Asiatic-Pacific Campaign Medal was awarded for service in the Asiatic-Pacific Theater of Operation from December 7, 1941 to March 2, 1946. This Theater included Alaska, Asia, Australia, New Zealand and the Pacific Islands. A 3/16th inch bronze star was awarded for each of the following 24 operations:

Aleutians	Marshall Islands
Asiatic - Pacific Raids - 1944	Minesweeping
Bismark Archipelago	New Georgia Group
Borneo	Netherlands East Indies
Eastern New Guinea	Okinawa Gunto
Escort, Antisubmarine & Armed Guard	Pacific Raids -1942
Iowa Jima	Pacific Raids -1943
Kurile Islands	Solomon Islands
Leyete	Special Operations
Luzon	Treasury - Bougainville
Manila Bay	Western Caroline Islands
Marianas	Western New Guinea

Many of these foregoing, major headings had multiple sub-operations for which a battle star was awarded.

Asiatic-Pacific Campaign Medal obverse.

Asiatic-Pacific Campaign Medal reverse.

European-African-Middle Eastern Campaign Medal

This medal known as the "ETO Medal" was awarded for service in the European Theater of Operations from December 7, 1941 to November 8, 1945. The ETO included Greenland, Europe, Africa and the Middle East. A 3/16th inch bronze star was issued for each of the 14 following operations:

Algeria-Morocco	Malta
Anzio-Nettuno	Normandy
Casablanca	Northeast Greenland
Elba and Pianosa	Salerno
Escort, antisubmarine, armed guard and special operations	Sicilian Occupation
Formia-Anzio	Southern France
Italy	Tunisia

ETO Campaign Medal obverse.

ETO Campaign Medal reverse.

USS Indiana (BB-58), descendant of one of America's original battleships, at sea, circa 1942. National Archives.

World War II Victory Medal

With U.S. withdrawal from the world stage, following the conclusion of World War I, a vacuum of power and political unrest was created, providing a ripe environment for despots to fill the void. Stalin rose to power in communist Russia and Hitler ascended to dictatorship in Germany. Lesser despots like Mussolini rose to power in Italy and Hideiki Tojo in Japan. Following preliminary moves in 1936, Hitler's war machine began to roll over Europe in 1939. By 1940 only a valiant British air corps, staffed in part by many volunteer U.S. pilots and crew members, stood between Hitler and the conquest of Great Britain.

At the same time, Japan, viewing the move of the U.S. Pacific fleet from San Francisco to Honolulu as a threat to its plans for conquest, signed a three party mutual defense agreement with Germany and Italy. The Japanese invasion of Indochina in early 1941 resulted in harsh economic sanctions from the U.S. When Japan launched a devastating surprise attack on Pearl Harbor on December 7, 1941, America's fate was sealed. The sneak attack had failed to destroy the sleeping tiger; instead it had unleashed its vengeance. America's entry into World War II would turn the tide in both Europe and the Pacific. Its industrial strength would contribute significantly to reversing the balance of power. The War would rage on for four more years and culminate with the dropping of two atom bombs on the Japanese mainland. In the end, only two super powers would remain: the U.S. and the Soviet Union, which would ultimately be revealed as a mere paper tiger.

The World War II Victory Medal was awarded to all of the more than 16 million members of the armed forces who served at least one day between December 7, 1941 and December 31, 1946 in any of the three Theaters of Operation (American, European-Middle Eastern-North Africa or Asiatic-Pacific). Of those who served more than 400,000 would be killed in action. The medal is unnumbered and incorporates a reference to the World War I Victory medal in its ribbon design.

World War II Victory Medal obverse.

World War II Victory Medal reverse.

Navy Occupation Service Medal

Established on January 28, 1948, this medal was authorized for the following periods of post-War World II occupation service:

Location	Clasp	Inclusive Dates
Austria	Europe	May 8, 1945 to October 25, 1955
Germany	Europe	May 8, 1945 to May 5, 1955
Italy	Europe	May 8, 1945 to December 15, 1947
Trieste	Europe	May 8, 1945 to October 25, 1955
Japan	Asia	September 2, 1945 to April 27, 1952
Korea	Asia	September 2, 1945 to April 27, 1952

Two clasps, "Europe" and "Asia," as appropriate, were authorized for wear on the suspension ribbon of the Navy Occupation Service Medal. This medal marks a return to the use of the distinctive Navy and distinctive Marine Corps reverses (straight "For Service"). It is also the last campaign medal to have been produced with such distinctive reverses. Henceforth, all campaign medals have shared a common reverse, specific to each medal. Each subsequent campaign medal reverse has been designed to be harmonious with the obverse of the medal.

Navy Occupation Service Medal obverse with two authorized Navy clasps: "Asia" and "Europe."

Navy Occupation Service Medal reverse. This is the last medal struck with a distinctive Navy reverse.

Humane Action Medal

On July 29, 1949 Congress authorized the presentation of this medal to all branches of the service participating in the Berlin airlift. The airlift was necessitated by the post-war ground and rail blockade of West Berlin by the Soviets from June 1948 to May 12, 1949. Nearly 300,000 flights landed at Templehof Airport to bring needed food and medical supplies to the West Berliners. This continued until the Soviets eventually capitulated and abandoned their attempts to isolate West Berlin for the moment.

Humane Action Medal obverse.

Humane Action Medal reverse.

4

POST WORLD WAR II

Korean Service Medal

At the conclusion of World War II, Korea was divided at the 38th parallel. In 1948, Syngman Rhee proclaimed the Republic of Korea south of the 38th parallel. His counterpart, Kim Il-Sung followed suit and proclaimed the area north of the 38th parallel as the Democratic Republic of Korea.

On June 25, 1950, North Korean forces invaded South Korea. The United Nations Security Council condemned the North Korean action and appointed General Douglas MacArthur to command the U.N. forces which would come to the aid of South Korea. By the end of 1950, U.N. forces had repulsed the North Korean invasion and invaded North Korea. When Communist China entered the fray in late 1951, U.N. forces were driven back from North Korean territory. Bitter fighting ensued for the next six months until a ceasefire was arranged in June of 1951. After a brief resumption of hostilities, a stalemate was reached in late 1951. Skirmishes, sometimes heavy, continued into 1953. On July 27, 1953 a truce was finally signed at Panmunjoun and the war was over. Status quo fractured by intermittent acts of aggression has continued for more than four decades.

The Korean Service Medal was authorized on November 8, 1950 for service between June 27, 1950 and July 27, 1954 for all military personnel participating in this war for a period of 30 consecutive or 60 non-consecutive days.

Bronze campaign stars were authorized for the following ten engagements:

Engagement	Date(s)
North Korean Aggression	1950
Communist Chinese Aggression	1950-51
Inchon Landing	1950
First U.N. Counteroffensive	1951
Communist Chinese Spring Offensive	1951
U.N. Summer/Fall Offensive	1951
Second Korean Winter	1951-52
Korean Defensive	1952
Third Korean Winter	1952-53
Korean Summer/Fall	1953

This conflict also marks a trend in the awarding of not only a U.S. campaign medal, but also a second medal from another organization (U.N., NATO, etc.) or country. In this case, the U.N. awarded a campaign medal in each of the following 12 languages (see Chapter VII):

Amharic	English	French	Italian	Spanish	Thai
Dutch	Flemish	Greek	Korean	Tagalog	Turkish

Korean Service Medal obverse. Note: 3/16th inch battle stars on suspension ribbon.

Korean Service Medal reverse.

National Defense Service Medal

The National Defense Service Medal continued the trend started during World War II of having a single medal for all branches of the service. Authorized on April 22, 1953, this medal has been awarded to all servicemen in all branches of service on active duty during any of the following three periods:

Period	Inclusive Dates	Action
1	June 27, 1950 to July 27, 1959	Korean Conflict
2	January 1, 1961 to August 14, 1974	Vietnam War
3.	August 2, 1990 to November 30, 1995	Operation Desert Storm

Additional awards of this medal are recognized by addition of a 3/16th inch bronze star to the suspension ribbon.

Due to the length of the second period of award, crude versions of the medal were manufactured in South Vietnam. This version is easily distinguished from the American original by the coarseness of the suspension ribbon and the crudeness of the planchet. These copies have assumed a value in their own right as curiosity pieces. There are also Japanese copies of this medal, but they are much less common than the South Vietnamese copies.

National Defense Service Medal obverse. Note: 3/16th inch bronze star denoting second award.

National Defense Service Medal reverse.

South Vietnamese made National Defense Service Medal obverse.

South Vietnamese made National Defense Service Medal reverse.

Antarctic Service Medal

Established on July 7, 1960, the Antarctic Service Medal was liberally awarded retroactively to 1946 for any period of service in Antarctica or its contiguous waters. Subsequent to June 1973, a minimum period of 30 days of service south of 60 degrees latitude was required. A "Wintered Over" clasp is awarded to those who have spent the winter months (March through October) in Antarctica. Bronze signifies one winter; gold two and silver three or more winters.

Antarctic Service Medal obverse.

Antarctic Service Medal reverse.

Antarctic Service Medal with bronze Wintered Over clasp signifying one winter spent on the continent.

Armed Forces Expeditionary Medal

Established on December 4, 1961, this medal is awarded for 30 days of service (or for the total period required, if less than 30 days) in support of the following types of operations: (1) U.S. operations; (2) U.N. operations; and (3) operations in support of U.S. allies. In certain hazardous operations, the time requirement is waived. A 3/16th inch bronze star is used to denote subsequent awards of this medal. A 3/16th inch silver star denotes receipt of 5 bronze stars. The 24 operations which have qualified for award of this medal include:

Operation/Area	Inclusive Dates
Lebanon	July 1, 1958 to November 1, 1958
Vietnam/Thailand	July 1, 1958 to July 3, 1965
Quemoy/Matsu	August 22, 1958 to June 1, 1963
Taiwan Straits	August 23, 1958 to January 1, 1959
Congo	July 14, 1960 to September 1, 1962
Laos	April 19, 1961 to October 7, 1962
Berlin	August 14, 1961 to June 1, 1963
Cuba	October 24, 1962 to June 1, 1963
Congo	November 23, 1964 to November 27, 1964
Dominican Republic	April 28, 1965 to September 21, 1966
Korea	October 1, 1966 to June 30, 1974
Cambodia	March 29, 1973 to August 15, 1973
Thailand	March 29, 1973 to August 15, 1973
Cambodia	April 11, 1975 to April 13, 1975
Vietnam	April 29, 1975 to April 30, 1975
Mayaguez Rescue	May 15, 1975 to May 15, 1975
Lebanon	June 1, 1983 to December 1, 1987
Grenada	October 23, 1983 to November 21, 1983
Libya	April 12, 1986 to April 17, 1986
Persian Gulf	July 24, 1987 to August 1, 1990
Panama	December 20, 1989 to January 31, 1990
Somalia	December 5, 1992 to March 31, 1995
Haiti	September 16, 1994 to March 31, 1995
S. W. Asia	December 1, 1995 to Date to be Determined

Armed Forces Expeditionary Medal obverse.

Armed Forces Expeditionary Medal reverse.

Vietnam Service Medal

Following the defeat of the French Colonial forces in L'Guerre Indochine, Vietnam was divided into the Democratic Republic of Vietnam (north) and the Republic of Vietnam (south) at the 1954 Geneva Peace Conference. By 1956, Hanoi began organizing resistance cells which remained in the South in violation of the Geneva agreements. By 1959, this activity had escalated into limited guerrilla warfare. The People's Army of North Vietnam's famed Group 559 began organizing the logistical flow of men and material down the Ho Chi Minh Trail. Group 559 would continue to play a pivotal role throughout the war.

During the 1960s, the American presence began to escalate, changing during the middle of the decade from an advisory role to the actual prosecution of the war. More than 500,000 men would be in country by 1969. While the Americans were ultimately able to destroy the Viet Cong infrastructure following the 1968 Tet Counteroffensive, they were at best only able to stave off the determined North Vietnamese efforts.

In early 1973, following intense bombing of the North throughout 1972 (Operations Linebacker and Linebacker II), a peace accord was signed in Paris. Americans left the field of combat during 1973 and the South Vietnamese were on their own. In early 1975, the North Vietnamese launched a Spring Offensive (code named Campaign 275). The Central Highlands of South Vietnam fell so quickly that even the North Vietnamese were surprised. After regrouping, they captured the coastal zones eventually attacking Saigon from three sides. On April 30, 1975 North Vietnamese tanks flying the Viet Cong flag crashed through the gates of the South Vietnamese Presidential Palace and the Second Indochina War was over.

The Vietnam Campaign Medal is considered by many to be one of the most attractive U.S. campaign medals ever designed. Its ribbon reflects the bright colors of the Republic of Vietnam's flag and the planchet's design incorporates a highly symbolic dragon behind a stand of bamboo leaves. The dragon, in this case, represents South Vietnam. This medal also served as an inspiration for the El Oro design of the Philippines Vietnam Service Medal. As with many of its post World War II predecessors, it was liberally awarded to all branches of the service. Vietnamese and Japanese copies of this medal exist. The South Vietnamese copies are crudely produced and usually have a uniface planchet with the dragon placed in front of the bamboo leaves. The suspension ribbon is also made of a very coarse weave. The Japanese copies are of a much higher quality and bear a closer, but still discernible, resemblance to the U.S. made medals.

Authorized on July 8, 1965, the Vietnam Service medal was awarded to all members of the armed forces who served in country or on the adjacent waters or over the country's airspace between July 3, 1965 and March 28, 1973. In addition, certain service in Cambodia, Laos and Thailand could also qualify for the award of this medal. As little as one day of service was required to qualify for award of this medal. The Navy recognized the following 17 distinct periods for award of a 3/16th inch battle star to be worn on the campaign medal:

Engagement	Inclusive Dates
Advisory Campaign	March 15, 1962 to March 7, 1965
Defensive Campaign	March 8, 1965 to December 24, 1965
Counteroffensive	December 25, 1965 to June 30, 1966
Counteroffensive II	July 1, 1966 to May 31, 1967
Counteroffensive III	June 1, 1967 to January 29, 1968
Tet Counteroffensive	January 30, 1968 to April 1, 1968
Counteroffensive IV	April 2, 1968 to June 30, 1968
Counteroffensive V	July 1, 1968 to November 1, 1968
Counteroffensive VI	November 2, 1968 to February 2, 1969
Tet 1969 Counteroffensive	February 23, 1969 to June 8, 1969
Summer/Fall 1969 Campaign	June 9, 1969 to October 31, 1969
Winter/Spring 1970 Campaign	November 1, 1969 to April 30, 1970
Sanctuary Counteroffensive	May 1, 1970 to June 30, 1970
Counteroffensive VII	July 1, 1970 to June 30, 1971
Consolidation I	July 1, 1971 to November 30, 1971
Consolidation II	December 1, 1971 to March 29, 1972
Ceasefire	March 30, 1972 to January 28, 1973

Continuing the trend started in Korea, the Vietnam Campaign Medal was accompanied by presentation of the Republic of Vietnam Campaign Medal (see Chapter VII).

Vietnam Service Medal obverse with 3 battle stars.

Vietnam Service Medal reverse.

RADM James B. Stockdale, who was awarded the Medal of Honor for the courage he displayed while a POW in North Vietnam delivers an address after his release. Stockdale is seen here wearing (top row left to right) the Legion of Merit, Distinguished Flying Cross, Air Medal, Purple Heart, American Campaign Medal, WW II Victory Medal and (bottom row left to right) Navy Occupation Service Medal, National Defense Service Medal, POW Medal, Vietnam Service Medal and the Republic of Vietnam Campaign Medal, circa 1974. National Archives.

South Vietnamese crudely produced Vietnam Service Medal obverse. Note: dragon is in front of bamboo stand. The suspension ribbon is also much more coarse than its U.S. counterpart.

Unfinished reverse of the South Vietnamese made Vietnam Service Medal.

Humanitarian Service Medal

This campaign medal is considered by many collectors and recipients alike to be the ugliest (by design) of all the campaign medals. Reaction most often centers on the single, open hand shown palm up, but others cite the garish colors of the suspension ribbon. Regardless, the combination certainly constitutes one of the weakest design efforts ever.

The Humanitarian Service Medal was established on April 1, 1975 to recognize direct participation in specified humanitarian efforts. To date, more than 100 foreign and domestic operations have been designated as qualifying for award of this medal. These operations have included assistance with both domestic as well as international relief efforts, evacuations, hostage rescues and assistance with natural disasters such as earthquakes and hurricanes.

Humanitarian Service Medal obverse.

Humanitarian Service Medal reverse.

Southwest Asia Service Medal

Established on March 15, 1991, the Southwest Asia Service Medal was awarded to all branches of the service for participation in the Middle East during Operations Desert Shield/Desert Storm. Battle stars (3/16th inch bronze stars) were authorized for the following three engagement periods:

Engagement	Inclusive Dates
Defense of Saudi Arabia	August 2, 1990 to January 16, 1991
Liberation & Defense of Kuwait	January 17, 1991 to April 11, 1992
Southwest Asia Ceasefire	April 12, 1991 to November 30, 1995

The award of this medal was usually accompanied by presentation of either the Kuwait Medal for the Liberation of Kuwait and/or the Saudi Arabia Medal for the Liberation of Kuwait (see Chapter VII).

Southwest Asia Service Medal obverse.

Southwest Asia Service Medal reverse.

Outstanding Volunteer Service Medal

This medal, which was authorized on January 9, 1993, is intended to recognize volunteer service of a civilian nature, i.e. outstanding community service reflecting sustained, individual commitment and involvement over a period of time. It is a reflection of our current shift to a peace time armed forces.

Outstanding Volunteer Service Medal obverse.

Outstanding Volunteer Service Medal reverse.

Armed Forces Service Medal

This is the newest campaign medal, authorized on January 11, 1996 for all branches of the service participating in significant military operations on or after June 1, 1992 in which no foreign opposition or imminent hostile action is encountered. Bosnia will be the first such operation for which award of the Armed Forces Medal will be made. It may also be accompanied by award of the NATO Medal for Bosnia (see Chapter VII) under certain conditions.

Armed Forces Service Medal obverse.

Armed Forces Service Medal reverse.

5

THE COMMEMORATIVE MEDALS

The Manila Bay Medal

On April 25, 1898 Congress, with the urging of President William McKinley, declared war on Spain. Commodore (promoted in 1899 to the rank of Admiral of the Navy for life by a grateful Congress) George Dewey, commander of the Asiatic Pacific fleet set sail from Hong Kong on his flagship Olympia. Along with the Raleigh, Boston, McCulloch, Concord, Petrel and Baltimore, they headed for

The original battleship USS Indiana (BB-1) passes in review in New York harbor following America's victory in the Spanish American War, circa 1898. Naval Historical Center.

Continued on page 53

Manila Bay. On May 1, 1898, the fleet entered Manila Bay and pounded the Spanish Pacific Squadron at Cavite into submission. The lives of only seven Americans would be lost in the battle. By sunrise, the Americans were in control of Manila Bay and for all practical purposes the entire Philippine Islands. Although intended only to shock Spain into compliance with American demands over Cuba, Dewey's actions would lead to the eventual annexation of the Philippines by the U.S.

In recognition of this stunning victory, Congress authorized the striking of 1,800 bronze medals by Tiffany & Co. in New York bearing the likeness of Commodore Dewey for presentation to the crews of the aforementioned ships. The medals, designed by Daniel Chester French, were each rim named, i.e. the recipient's name and rank was stamped on the medal's edge. The ship's name is stamped on the reverse. These medals reflect the intricacy of fine craftsmanship. They also use a very unusual brooch design. Unnamed copies of this medal exist and originals are encountered with replacement ribbons. The issuance of this medal predates that of any of the Navy campaign medals.

Incidentally, Dewey, himself, always wore his medal with the reverse facing outwards out of a sense of modesty.

Manila Bay (Dewey) Medal obverse.

Manila Bay Medal: detail of reverse.

The Sampson Medal

This commemorative medal was authorized by Congress in 1901 to recognize the Navy's victories in the Caribbean during the Spanish American War. It is often referred to as the West Indies or West Indies "98" Campaign Medal creating some confusion with the actual Navy West Indies Campaign Medal, discussed previously in Chapter I. The medal is actually a commemorative medal featuring the likeness of Admiral William T. Sampson on the planchet's obverse. Sampson earlier chaired the investigation into the sinking of the Battleship Maine. He was later promoted to Commodore and then Rear Admiral. He was subsequently appointed by then Secretary of the Navy John D. Long to command the North Atlantic Squadron in 1898 during the Spanish American War.

Sampson was credited with the defeat of the Spanish squadron in the Caribbean during the war. The action began with a courageous, but unsuccessful attempt by Navy Lieutenant Richmond P. Hobson and a volunteer crew of seven to sink the coal carrier Merrimac inside a narrow strait. Their intended objective was to bottle the Spanish squadron inside Santiago harbor for the duration of the war. Admiral Sampson subsequently ordered a major ground attack by 17,000 Army troops against the city. When Spanish Rear Admiral Pascual Cervera y Topete's ships were ordered to escape from Santiago Harbor at all cost by the Spanish Governor on July 3, 1898, they ran straight into the heavy guns of Sampson's ships. The Spanish ships were destroyed in a quick one sided victory reminiscent of Dewey's triumph in Manila Bay.

The original Type I medal used an unusual brooch bearing the name of the recipient's ship. From this top bar, engagement bars were attached and used to suspend the medal. In 1908, the medal (Type II) was redesigned so that engagement clasps were worn below the ship's name bar, similar to the clasps for the World War I Victory Medal, discussed previously in Chapter I. The original engagement for which the medal was awarded is engraved on the reverse.

The crews of the following 67 ships (top bars bearing the names of these ships exist) qualified for award of the Sampson Commemorative Medal. The list also reflects the initial action for which the medal was awarded, i.e. the engagement which was engraved on the reverse of the medal's planchet:

Continued on page 54

Ship	Engagement	Ship	Engagement
Abarenda	Guantanamo	Morrill	Havana
Alvardo	Manzanillo	Nashville	Cienfuegos
Amphitrite	San Juan	Newark	Santiago
Annapolis	Barcoa	New Orleans	Santiago
Bancroft	Cortes Bay	New York	Santiago
Brooklyn	Santiago	Oregon	Santiago
Cincinnati	Matanzas	Osceola	Manzanillo
Castine	Mariel	Panther	Guantanamo
Detroit	San Juan	Peoria	Tunas
Dixie	Casilda	Porter	San Juan
Dolphin	Santiago	Prairie	Mariel
Dupont	Matanzas	Puritan	Matanzas
Eagle	Cienfuegos	Resolute	Santiago
Ericsson	Santiago	St. Louis	Santiago
Fern	Santiago	St. Paul	San Juan
Gloucester	Santiago	Suwanee	Santiago
Hamilton	Mariel	San Francisco	Havana
Harvard	Santiago	Scorpion	Manzanillo
Hawk	Mariel	Terror	San Juan
Helena	Tunas	Texas	Santiago
Hist	Santiago	Topeka	Nipe Bay
Hornet	Manzanillo	Vixen	Santiago
Indiana	Santiago	Vesuvius	Santiago
Iowa	Santiago	Vicksburg	Havana
Leyden	Nipe Bay	Wasp	Cabanas
Machias	Cardenas	Wilmington	Cardenas
Mangrove	Caibairien	Windom	Cienfuegos
Manning	Cabanas	Winslow	Cardenas
Maple	Isle of Pines	Wompatuck	Santiago
Marblehead	Cienfuegos	Yale	San Juan
Massachusetts	Santiago	Yankton	Cape Muno
Mayflower	Havana	Yankee	Santiago
McKee	Sagua la Grande	Yosemite	San Juan
Montgomery	San Juan		

Only the following 30 ships qualified for additional engagement clasps, i.e. clasps which are mounted on the suspension ribbon of the medal's Type II version just below the ship's name:

Ship	Engagement(s)
Bancroft	Isle of Pines
Brooklyn	Santiago
Dolphin	Guantanamo
Dupont	Santiago
Eagle	Cape Muno
Gloucester	Santiago
Helena	Manzanillo, Tunas
Hist	Manzanillo
Hornet	Manzanillo
Indiana	San Juan, Santiago
Iowa	San Juan, Santiago
Manning	Mariel, Naguerro
Marblehead	Cienfuegos, Gunatanamo, Santiago
Massachusetts	Santiago
Newark	Manzanillo
New Orleans	Santiago
New York	Matanzas, San Juan, Santiago
Oregon	Santiago
Osceola	Manzanillo, Tunas
Porter	Santiago
Resolute	Manzanillo
Suwanee	Guantanamo, Manzanillo
Scorpion	Manzanillo
Texas	Guantanamo, Santiago
Vixen	Santiago
Vesuvius	Santiago
Wasp	Mariel, Nipe Bay
Wilmington	Manzanillo
Wompatuck	Manzanillo
Yankee	Casilda, Cienfuegos

The following 21 distinctive engagement clasps (in addition to the engagement for which the medal was awarded) with the date of each additional engagement engraved on the reverse of each clasp were authorized for display on the Sampson Medal. All locations are in Cuba, except San Juan, which is in Puerto Rico.

Barcoa	Caibarien	Matanzas	Rio Hondo
Casilda	Guantanamo	Mariel	Sagua al Grande
Cape Muno	Havana	Nipe Bay	San Juan
Cabanas	Isle of Pines	Naguerro	Santiago
Cardenas	Manzanillo	Punta Colorado	Tunas
Cienfuegos			

Clearly, the more obscure clasps reflect the higher values especially when combined with the more obscure ships. In addition, the original of these medals are rim named as is the case with the Manila Bay Medal. Conventional wisdom would place higher values on the smaller ships and those engaged in significant battles. Higher values would also be assigned to medals named to officers. Original issue, unnamed medals also exist as well as restrikes. The current restrike does not utilize the ship's name on the brooch.

Sampson Medal Type II obverse.

Sampson Medal: detail of ship's name bar.

Sampson Medal: detail of reverse.

Seaman Willard Dwight Miller, a Canadian, who along with his brother, Harry, won the Medal of Honor at Cienfuegos, Cuba for heroism during the cutting of the trans-Atlantic cable on May 11, 1898. He is also wearing the Sampson Commemorative Medal (middle) and the Navy Good Conduct Medal (right). Naval Historical Center.

Specially Meritorious Medal

Authorized on March 3, 1901, this medal was presented to 93 recipients for the rescue of crews from burning ships following the naval battle at Santiago, Cuba on July 3, 1898. These actions followed the battle which erupted when the Spanish Governor of Cuba ordered Rear Admiral Pascual Cervera y Topete's four fast cruisers to escape from Santiago Harbor at any cost. Admiral Cervera's ships were engaged by the U.S. ships Brooklyn, Indiana, Oregon, Iowa and Texas as they tried to flee Santiago Harbor. The U.S. ships soon overwhelmed the Spanish. The flagship Maria Teresa was run aground as were the Spanish cruisers Oquendo and Vizcaya. In a separate action the converted steam yacht, Gloucester, under the command of Lieutenant Commander Richard Wainwright, who incidentally was not a recipient of the Specially Meritorious Medal, sunk a Spanish destroyer.

The medal itself is a smallish 1.25 inch cross pattee (see glossary) with an engraved reverse. The reverse is inscribed with the recipient's name and rank along with the date (July 3, 1898) and the location of the action (Santiago, Cuba). In two other cases, the medals awarded to Navy LT. Richmond Hobson, who sank the Merrimac inside the harbor at Santiago and Navy LT. Victor Blue bear different inscriptions on the reverse. The reverse of Hobson's and Blue's medals read: "...sinking the Merrimac in entrance to Santiago Harbor, June 3, 1898" and "...locating the enemy's ships in Santiago Harbor, June 12, 1898," respectively. In 1933, LT. Hobson was also awarded the Congressional Medal of Honor for his actions at Santiago.

In addition to the 93 recipients named in 1904, Strandberg and Bender report that there were an additional eight "unofficial" recipients, who seem to have legitimately qualified for award of this medal. The six ships involved in this action and the number of recipients from each include:

Ship	Recipients
Ericsson	22
Gloucester	32
Harvard	8
Hist	9
Indiana	5
Iowa	13

There are also original strikes marked "Display" or "Exhibition" on the reverse. Darker bronze copies (fakes) of this medal also exist in limited numbers.

Specially Meritorious Medal obverse.

Cardenas Medal of Honor

This is the fourth in the series of commemorative medals emanating from the 1898 Spanish American War. It was authorized by Congress on May 3, 1900 for presentation exclusively to the officers and crew of the Revenue Cutter Hudson for gallantry in action while towing another ship, the Winslow, to safety during the battle at Cardenas, Cuba on May 11, 1898. Although the Revenue Cutter Service was a forerunner of the Coast Guard, this medal is included here because the Hudson was operating at Cardenas under operational control of Admiral Sampson. The Winslow, which had become disabled, was facing annihilation from the Spanish shore batteries. Braving heavy fire, the men of the Hudson managed to make fast a line to the Winslow and tow it out of the range of the enemy's guns. Originally struck as a "table top" medal (one not intended for wear on the uniform), it was later converted to a wearable medal. It was presented in gold to Frank Newcomb, Captain of the Hudson, in silver to the officers and bronze to the enlisted men.

Bronze Cardenas Medal of Honor Table Top Version Reverse.

Bronze Cardenas Medal of Honor Table Top Version obverse.

Bronze Cardenas Medal of Honor Wearing Version Obverse.

Bronze Cardenas Medal of Honor Wearing Version Reverse. All Cardenas Medal of Honor photos by PAC David Santos, USCG made available courtesy of Cindee Herrick, Curator, U.S. Coast Guard Museum, New London, CT.

The NC-4 Medal

This commemorative medal was authorized on February 9, 1929. It was originally presented as a "table top" medal to Lieutenant Commander Albert C. Read (the pilot) and the five crew members of the Navy Curtis-4 flying boat for successful completion of the first transatlantic flight on May 8, 1919 from Nova Scotia to the Azores. A medal was also authorized for Commander John H. Towers, Division Commander. In 1935, a wearable version of the medal was created. The wearable version contains the names of all recipients on the reverse of each medal. The seven recipients were:

CDR John H. Towers
LCDR Albert C. Read

LT James L. Breeze
LT Walter Hinton
LT Elmer F. Stone
ENS Herbert C. Rodd, and
CMM Eugene S. Rhodes

Little known to most at the time, Richard E. Byrd played a key role in developing the navigation techniques used by the NC-4 crew during their successful flight. Such techniques were to play a key role in Byrd's later aerial explorations of both the North and South Poles (see Chapter III). Limited copies (fakes) of the wearable medal are known to exist.

NC-4 Wearing Medal obverse.

NC-4 Wearing Medal reverse. Note the inscription of the names of all recipients as an integral part of the design of this medal's reverse.

NC-4 Table Top Medal. The Navy Museum, Washington, D.C.

6

THE ARCTIC/ANTARCTIC MEDALS

Admiral Richard E. Byrd, famed Naval navigator and Polar explorer, circa 1925. National Archives.

The history of our country's polar explorations is inextricably tied to the efforts of the U.S. Navy. The medals discussed in this Chapter are of a commemorative nature and the stories surrounding the presentation of each are the tales of legend. For this reason, coverage of these extremely rare medals is included in this work.

The Jeannette Arctic Expedition Medal

August Peterman, editor of the *Georgrapische Mitteilunggen,* popularized the theory of the "Open Polar Sea" during the mid-19th century. One of his strongest adherents was Navy Commander George Washington DeLong (1844-1881). Peterman's theory held that a passage through the otherwise apparently impenetrable ice field surrounding the North Pole led to an unfrozen, open sea.

In 1873, DeLong made his first journey to the Arctic regions along the coast of Greenland to search for the missing Charles Francis Hall Expedition (1871-1873) and their ship, *Polaris.* Upon his return from the unsuccessful effort, DeLong contacted James Gordon Bennett, proprietor of the *New York Herald,* to drum up financial support for an Arctic expedition of his own. At a time when exclusive newspaper coverage of such events generated higher sales, DeLong's task was not too difficult. With Bennett's financial backing, DeLong traveled to England to purchase the 146 foot-long Arctic exploration vessel, Pandora, which was rechristened Jeannette in honor of Bennett's sister.

The refitted Jeannette departed San Francisco on July 8, 1879, after a stopover to replenish her supplies. Dubbed the "American Arctic Expedition of 1879," the project team was led by naval officers operating under military regulations, staffed by civilians and funded by Bennett. In addition to the 35 year old DeLong, the other naval officers were Charles William Chipp and John Danenhower.

The Jeannette reached the Bering Straits by the end of August 1879. By early September, the ship had reached the vicinity of Herald Island, the expedition's "wintering over" spot. Before actually reaching landfall, the ship was shrouded in fog and soon became frozen fast in the ice. The Jeannette then began an erratic triangular drift which carried her far to the west of her original position. This situation continued unabated for two years, leaving the shipbound crew to wonder if the ice would ever set them free. Fortunately, the expedition had been outfitted with sufficient food supplies for two years, supplemented by the bounty from occasional hunting trips on the ice by the Expedition's two Eskimos, Alexey and Aniguin, to withstand this situation.

When the ice around the ship began to loosen in early June of 1881, spirits were lifted, but the euphoria would be short lived. On June 11th, a strong Arctic wind jammed the ice hard along the ship's port side cracking her timbers. The ship began to slowly sink. Reacting with stoic professionalism, DeLong ordered the whale boats and supplies onto the ice pack. By the next day the Jeannette had gone under.

With adequate supplies, the Expedition made for the Siberian mainland. When they finally reached open water, the members of the Expedition and the supplies were divided into three separate groups. The groups southward journey continued without event until September 12. During that afternoon wind speeds began increasing. By 9 pm the three boats had become separated and would never again join together. One of the boats, under the command of LT Chipp, carrying seven other members of the Expedition, including Medal of Honor nominee for his heroics on the Expedition, Alfred Sweetman, disappeared without any further trace. Another boat, under the command of Chief Engineer, George Melville, eventually reached the Lena River Delta. All 11 men would be rescued by local hunters and eventually return to the United States.

The cruelest fate, however awaited DeLong and the 13 men with him. Reaching the northern most point of the Lena River Delta, they would be more than 100 miles from safety. By September 22, DeLong's party began to weaken. Food stocks began to dwindle to a dangerous level by early October. On the 6th, Seaman Hans Ericksen, who five days earlier had five toes amputated by Dr. James Ambler, died. His body, wrapped in canvas, was placed on the frozen river. The party was too weak to dig a grave in the frozen ground. The two strongest members of the group, William Nindemann, who was also nominated by DeLong for the Medal of Honor for his heroics, and Louis Noros were dispatched by De Long to seek help.

On October 30, 140 days after leaving the Jeannette, only DeLong and one of the two Chinese cooks, Ah Sam, were still alive. Ironically, Nindemann and Noros were able to locate help, but were not able to overcome the language barrier between themselves and the Tungus, native Siberians who managed a meager existence along the Lena River. When they finally joined Melville's group all hope had been lost for DeLong. His body would be found by Melville and Nindemann on May 6, 1882. His meticulously kept notes would be found near his body. In all, 20 men of the original 33 had perished under the most difficult circumstances. The publication of DeLong's meticulously kept log, which was preserved in a copper cylinder found near his body, and the eventual return of the bodies of his party to the United States in 1883 restored a measure of dignity to their effort.

In September, 1890, reacting to very strong public pressure, Congress authorized the production of 33 commemorative medals. Another 17 months would pass before the Secretary of the Navy would approve the design of the medal. Eight gold medals would be presented to the officers and senior civilian members of the Expedition. An additional 25 medals were struck in silver for the remaining members of the Expedition. The most unusual aspect of this medal's design is the suspension brooch. The brooch features an eagle holding the ship's name in its talons. Each recipient's name is engraved in curved style on the reverse.

For those wishing to view this medal, the silver medal awarded to the Chinese Mess Steward Ah Sam, who perished on the Lena Delta along with DeLong, is on display at the Navy Museum in Washington, D.C.

The Jeannette goes under June 12, 1881. The Navy Museum, Washington, D.C.

Jeannette Medal obverse. Courtesy Charles P. McDowell.

Jeannette Medal in plush velvet case of issue. Courtesy of Charles P. McDowell.

Jeannette Medal reverse named to Walter Sharvell. Courtesy Charles P. McDowell.

Peary Polar Expedition Medal

Robert Edwin Peary (1856-1920) was one of the most obsessive men to ever seek polar glory. He began his quest for fame with a number of early explorations in Greenland. Frederick Cook, who in later years would become Peary's chief protagonist, served as surgeon on those early trips. Along with Peary on these early efforts was his constant companion and virtual servant, Matt Henson. On one of the early trips when Peary lost a few toes to frostbite, he was quoted as saying, "...a few toes aren't much to give to achieve the Pole." In 1898, while the Spanish American War was raging in the Atlantic and Pacific, Peary along with Henson reached 84 degrees, 17'27" north. In 1902 at 46 years of age he returned home feeling too old for any further attempts.

Nevertheless in 1906, Peary was back for another try. With Henson leading the way, he reached 87 degrees 6' north, a new record to date. In 1908, he returned for his final attempt. This time he was facing stiff competition from an old ally, Frederick Cook. Both would eventually claim the Pole within weeks of each other, a debate that continues to simmer to this very day.

On April 1, 1909 after reaching 87 degrees 47' north, Peary dismissed their four Eskimos (Ooqueah, Ootah, Egingwah and Seegloo) and he and Henson set off alone for the Pole. On April 6, they recorded 89 degrees 57' north. By 6 pm that evening, they claimed to have reached the Pole. On April 7, they turned for home

to soon learn of Cook's claim of Polar superiority. The debate raged like a fire storm with many discrepancies being poked in the stories of both parties. In part, Peary's claim would have required him to cover incredible distances over demanding and dangerous terrain in a very short period of time. Nevertheless, the powerful National Geographic Society supported Peary's claim.

This is all by way of background to indicate that the eventual authorization of an expedition medal was just another in a series of acts to legitimize Peary's claim. In the end, the medal was awarded to six of the 1908-1909 expedition's members, but not Peary, himself. This was an unbelievable repudiation. The six recipients included:

Captain Robert Abram Bartlett
George Borup (posthumously)
Dr. John Godsell
Matthew Henson
Commander Donald MacMillan
Ross Marvin (who drowned during the expedition)

Peary's claim has survived, but his bitterness and obsessiveness won him no friends. Well made copies (fakes) in limited numbers of the medal exist. These copies have taken on an exceptional value all to themselves.

Peary Polar Expedition Medal obverse.

Peary Polar Expedition Medal reverse.

Byrd First Antarctic Expedition

Richard Evelyn Byrd (1888-1957) was one of America's influential polar explorers. Following a successful (but disputed and subsequently discredited) attempt to fly over the North Pole with his pilot Navy Chief Machinist Floyd Bennett on May 9, 1926, Lieutenant Commander Byrd was hailed as a national hero. He received the Distinguished Service Cross, the Distinguished Flying Cross and the Hubbard Medal of the National Geographic Society, which was presented by then President Calvin Coolidge. Byrd's place in history as America's premier polar explorer was secured.

Byrd was later elevated to full Commander and he and Bennett were both awarded the Medal of Honor by a grateful Congress in 1927. He then turned his attention to the South Pole. Here his contributions to polar research have become legend. With the financial backing of rich and powerful supporters such as Edsel Ford and John D. Rockefeller, he was able to mount several successful explorations of the South Pole.

On the first Antarctic Expedition (1928 to 1930), Byrd established a base camp, which became known as Little America. Byrd's principal plane was a Ford Trimotor, named Floyd Bennett in honor of Byrd's Arctic companion and pilot. It was powered by a 525-horsepower Wright Cyclone nose engine and two 250-horsepower Wright Whirlwinds under each wing. Departing from Little America, Byrd, along with his chief pilot Bernt Balchen and crew members Harold June, and Ashley "Mac" McKinley, became the first to fly over the South Pole at 1:14 am on November 29, 1929. Although surrounded by the journalistic sensationalism, which seemed to accompany all "firsts" during the 20s, Byrd's expedition contributed a great deal to the scientific understanding of the region.

On June 20, 1930, Byrd and his family, as well as the other members of the expedition, were received at the White House by President Herbert Hoover. Later that evening Byrd was presented with a special gold Medal of Honor from the National Geographic Society. Byrd was also promoted to Rear Admiral on the Navy's retired list. In addition, the Byrd Antarctic Medal, which had been authorized by Congress on May 23, 1930 was presented to members of the expedition. A total of 82 finely engraved medals were issued: 66 gold, seven silver and nine bronze. High quality, limited copies (fakes) are known to exist.

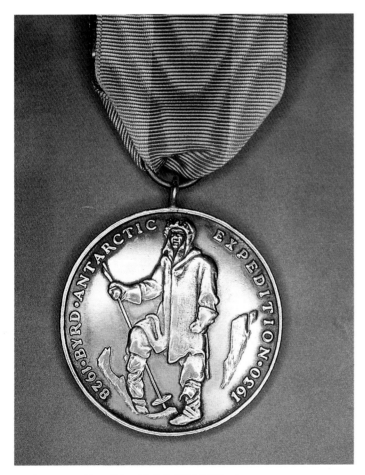

Byrd Antarctic Expedition Medal, 1928-1930, obverse.

Byrd Antarctic Expedition, 1928-1930, reverse.

Byrd Second Antarctic Expedition

In 1933 Byrd returned to Little America with private financial backing. The expedition would become noted for its live weekly radio broadcasts to the American public. Over the next two years, Byrd and the other members of the Expedition would conduct extremely valuable scientific research on the Antarctic Continent. Byrd, himself, would spend an entire winter in isolation at one of their advance bases.

Upon their return in 1935, Congress again moved to honor their efforts approving a new polar medal on June 2, 1936. In all, 57 silver medals were issued. High quality, limited copies (fakes) exist.

Byrd Antarctic Expedition, 1933-1935, obverse.

Byrd Antarctic Expedition, 1933-1935, reverse.

U.S. Antarctic Expedition Medal

In 1939, the United States Antarctic Service (USAS) was established using combined private and government funding. Admiral Byrd was appointed as the initial director of the USAS. The new service conducted only a single Polar expedition (1939 to 1941). They reestablished the base at Little America and a second base on the Antarctic Peninsula. Initially it had been hoped that the expedition would be an on-going effort. During his tenure, the Navy effectively eased Byrd out of operational control of the USAS and with the start of World War II ended all further efforts on the Antarctic continent. Byrd returned to active duty and spent much of the War mapping numerous Pacific Islands. He was later involved in the Navy's Operation Highjump at Little America in 1946, but was not in command.

Congress authorized a medal to commemorate the sole USAS Expedition on September 24, 1945. A total of 160 medals were created: 60 gold, 50 silver and 50 bronze. Admiral Byrd was a recipient of each of the three commemorative Antarctic medals.

U.S. Antarctic Service Expedition Medal obverse. Courtesy Charles P. McDowell.

Reverse of U.S. Antarctic Service Expedition Medal named to LCDR Clay W. Bailey, USN. Courtesy Charles P. McDowell.

Jeannette Medal presented posthumously to the Chinese Mess Steward Ah Sam and the copper cylinder containing George DeLong's meticulously kept log.

(LEFT) Admiral Robert E. Peary on 1909 Polar Expedition. National Archives.

7

RIM NUMBERING, NAMING, REVERSES AND BROOCH STYLES

Although discussed earlier, the significance of rim numbering, naming, reverses and brooch styles is to help the collector to identify authentic material. For this reason alone, it seems worthwhile to summarize the key factors of each of these three issues.

Rim Numbering

Prior to World War II, all Navy and Marine Corps campaign medals bore either a "M. No." prefixed number, indicating their source as the U.S. Mint, or a non-prefixed or plain number on the rim of the planchet, indicating a private or contract manufacturer. No Navy or Marine Corps campaign medals were issued with a "No." prefix. It is possible that a few medals may have been issued outside of the sequences reported below and would still be considered to be authentic.

Sketch of plain and "M. No." rim numbering. Courtesy Ned Broderick, National Vietnam Veterans Art Museum, Chicago.

"M. No." prefixed Navy medals include:

Medal	M. No. Sequence
Phillipine Campaign	4193 to 4392
Mexican Service	15500 to 16674
Second Nicaraguan Campaign	1 to 10000
Yangtze Service	1 to 13500

Non-prefixed Navy medals include:

Medal	Plain Number Sequence
Civil War Campaign	1 to 2,700
West Indies Campaign	1 to 3090
Spanish Campaign	1 to 6050
Philippine Campaign	1 to 4192
China Relief Expedition 1901	1 to 400
China Relief Expedition 1900	401 to 1150
Cuban Pacification	1 to 2100
First Nicaraguan Campaign	1 to 1500
Mexican Service	1 to 15499
Haitian Campaign (1915)	1 to 4000 and 4900 to 5300
Dominican Campaign	1 to 3800

"M. No." prefixed Marine Corps medals include:

Medal	M. No. Sequence
Second Nicaraguan Campaign	1 to 6000
Yangtze Service	1 to 6000
MC Expeditionary	1 to 10000

Non-prefixed Marine Corps medals include:

Medal	Plain Number Sequence
Civil War	1 to 200
West Indies Campaign	1 to 400
Spanish Campaign	401 to 900
Phillipine Campaign	1 to 1275
China Relief Expedition 1900	1 to 600
Cuban Pacification	1 to 1550
First Nicaraguan Campaign	1 to 1100
Mexican Service	1 to 2400
Dominican Campaign	1 to 2800
Haitian Campaign (1919-1920)	1 to 3250
China Service	1 to 4000

The West Indies Campaign and Spanish Campaign medals share the same numbering sequence. The Navy's non-prefixed Haitian Campaign (1915) Medal's numbers includes those medals issued to Marine Corp personnel, which are generally in the 3000 range. The size and style of the non-prefixed numbers can vary widely based on the manufacturer of the medal.

Curved Navy reverse.

Curved Marine Corps reverse.

Naming

Only two U.S. Navy Medals were officially rim named: the Manila Bay Medal and the Sampson Medal. Each original strike of these medals bore the recipients name and rank engraved on the rim. In the case of the Manila Bay Medal the recipient's ship's name is also engraved on the reverse. In the case of the Sampson Medal, the ship's name serves as the top bar and the initial engagement for which the medal was awarded is engraved on the reverse of the planchet. Other commemorative medals such as the Specially Meritorious Medal and the Jeannette Medal were "reverse" named, i.e. the recipient's name was engraved on the reverse of the planchet. In other cases, campaign medals have been engraved with the recipient's name on the rim as a result of personal action on the part of the recipient.

Reverses

As discussed in the Introduction, all Navy campaign medals and all Marine Corps campaign medals share a common reverse distinctive to each, with the exception of how the words "For Service" are displayed (curved or straight). The following table summarizes the appropriate format of these words for each campaign medal bearing a distinctive Navy or Marine Corps reverse:

Medal	Reverse
China Relief	Curved
China Service	Straight
Civil War	Curved
Cuban Pacification	Curved

Straight Navy reverse.

Straight Marine Corps reverse.

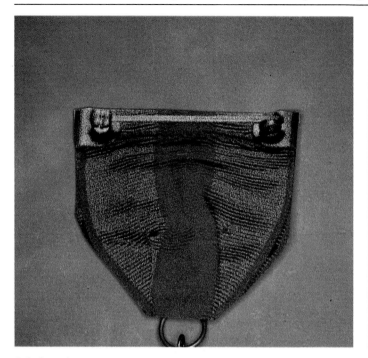

Split brooch.

Full wrap brooch.

Dominican	Straight
First Nicaraguan	Curved
Haitian 1915	Curved
Haitian 1919-1920	Curved
Mexican Service	Curved (1 to 15499) and Straight (M. No. 15500 to 16674)
MC Expeditionary	Straight
Navy Expeditionary	Straight
Navy Occupation Service	Straight

Philippine	Curved (1 to 4192) and Straight (M. No.4193 to 4392)
Second Nicaraguan	Straight
Spanish	Curved
West Indies	Curved
Yangtze Service	Straight

Note: Only the Marine Corps Expeditionary and the Navy Expeditionary Medals continue to be available for award and even so with

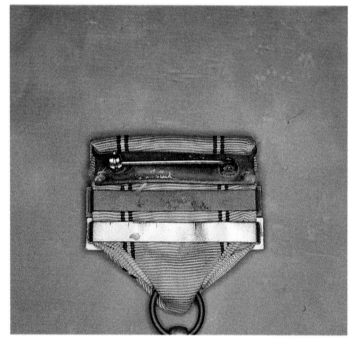

Slot brooch.

Crimp brooch.

Continued on page 69

less frequency due to the existence of the Armed Forces Expeditionary Medal.

Brooch Styles
There are basically four brooch styles: (1) full wrap, (2) split; (3) slot and (4) crimp. To a degree, the various brooch styles may be assigned to certain time periods. This can assist in identifying an original strike, but it is by no means a definitive method of determining an original medal. The following tables indicate the general period of usage of the various styles on Navy and Marine Corps medals and the style used most frequently with each medal:

Table I

Brooch Style	General Period
Split	1898 to 1932
Wrap	1933 to 1945
Slot	1946 to 1960
Crimp	1961 to Date

Table II

Style	Medal
Split	Civil War Campaign
	West Indies Campaign
	Spanish Campaign
	Philippine Campaign
	China Relief Expedition
	Nicaraguan Campaign
	Mexican Service
	Haitian Campaign (1915)
	Second Nicaraguan
Wrap	Dominican Campaign
	WW I Victory
	Haitian Campaign (1919-20)
	MC Expeditionary
	Yangtze Service
	Navy Expeditionary
	China Service
Slot	American Defense
	American Campaign
	European-African-Middle Eastern Campaign
	Asiatic-Pacific Campaign
	WW II Victory
	Navy Occupation
Crimp	Humane Action
	Korean Service
	National Defense Service
	Armed Forces Expeditionary
	Vietnam Service
	Southwest Asia Service
	Antarctica Service
	Humanitarian Service
	Outstanding Volunteer Service
	Armed Forces Service

Medals which have had their ribbons replaced will often be updated to a current brooch style. It is possible therefore to find an original medal with a current brooch.

8

MEDAL DEVICES

There are numerous devices which may be appended to the suspension ribbons of Navy and Marine Corps campaign medals. This chapter will focus on the more prevalent ones that a collector is likely to encounter.

Service/Battle Stars

Service or battle stars are utilized by all branches of the armed forces, primarily to denote subsequent awards of the same campaign medal or participation in the various recognized segments of a single campaign. The Navy and Marine Corps system is based on a 3/16th inch bronze star for the second through fourth award and a 3/16th inch silver star in lieu of five bronze stars. The following eight medals are broken down by authorized engagements or segments, each of which merits a service/battle stars:

Medal	Segments
Armed Forces Expeditionary Medal	24
China Service Medal	2
Humanitarian Service Medal	104
Korean Service Medal	10
Marine Corps Expeditionary Medal	66
National Defense Service Medal	3
Navy Expeditionary Medal	68
Southwest Asia Service Medal	3
Vietnam Service Medal	17

The following table summarizes the service/battle stars and devices other than clasps authorized for Navy and Marine Corps medals and service ribbons:

Medal/Ribbon	Bronze	Silver	Other Device
Armed Forces Expeditionary	Yes	Yes	MC
American Campaign	Yes	Yes	MC
American Defense	Yes	N/A	Block Style A
Armed Forces Service	N/A	N/A	N/A
Asiatic-Pacific Campaign	Yes	Yes	MC
China Service	Yes	N/A	N/A
European-African-Middle Eastern	Yes	Yes	MC
Humanitarian Service	Yes	Yes	N/A
Korean Service	Yes	Yes	MC
National Defense Service	Yes	N/A	N/A
Marine Corps Expeditionary	Yes	Yes	W

3/16th inch silver star denotes fifth award of Armed Forces Expeditionary Medal.

3/16th inch bronze star denotes second award of National Defense Service Medal.

Continued on page 71

1919-1920 clasp on First Haitian Campaign Medal (1915).

The rare Wake Island clasp on Navy Expeditionary Medal. The presence of this clasp greatly enhances the value of this medal.

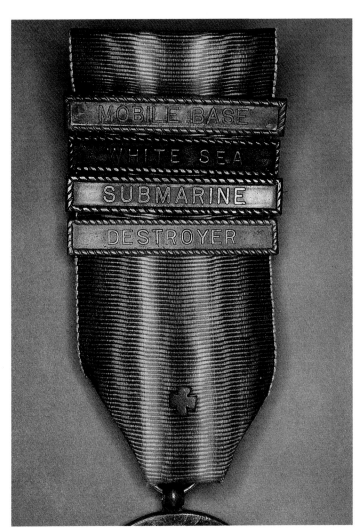

Meritorious Unit Commendation	Yes	Yes	N/A
Naval Reserve Sea Service	Yes	Yes	N/A
Navy Expeditionary	Yes	Yes	W
Navy Overseas	Yes	Yes	N/A
Navy Unit Commendation	Yes	Yes	N/A
Presidential Unit Citation	Yes	Yes	Globe,Gold N
Sea Service	Yes	Yes	N/A
Southwest Asia	Yes	N/A	MC
Vietnam Service	Yes	Yes	MC
Volunteer Service	Yes	N/A	N/A
WW I Victory Medal	N/A	N/A	Bronze Maltese Cross

Notes on the table: (1) The bronze globe was awarded to personnel of the submarine USS Triton when cited and the gold "N" was awarded to personnel of the submarine USS Nautilus for display

West Indies clasp mounted on WW I Victory Medal.

Mobile Base, White Sea, Submarine and Destroyer clasps mounted on WW I Victory Medal for display purposes only. Only a single Army or Navy service clasp was authorized for display on the suspension ribbon of the WW I Victory Medal.

Continued on page 72

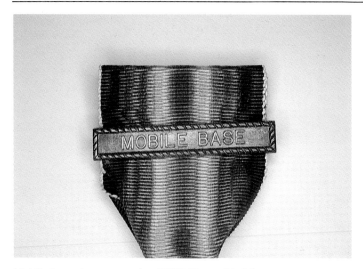

Mobile Base clasp mounted on WW I Victory Medal.

Fleet and Base clasps mounted on American Defense Medal.

on the Presidential Unit Citation service ribbon; (2) the silver "W" may be worn on the Navy Expeditionary service ribbon (and the Marine Corps Expeditionary service ribbon) by personnel who participated in the defense of Wake Island during WW II; (3) the bronze "A" is worn on the American Defense service ribbon by personnel of cited ships during the period immediately preceding WW II; (4) the bronze Marine Corps insignia (noted as "MC" in the above table) is worn on the applicable service ribbons, as noted, by Navy personnel assigned to Fleet Marine Force units and (5) the bronze Maltese Cross was authorized for wear by members of the American Expeditionary Force during WW I (it can only be worn with a service clasp).

Clasps

A number of Navy campaign medals may be worn with 1.5 by 1/4 inch rectangular clasps with stylized rope borders. The following table summarizes the various clasps assigned to each of these medals:

Medal	Authorized Clasp(s)
American Defense Service	Base
	Fleet
Antarctic Service	Wintered Over
Haitian Campaign 1915	1919-1920
MC Expeditionary	Wake Island
Navy Expeditionary	Wake Island
Navy Occupation Service	Asia
	Europe
World War I Victory	Armed Guard
	Asiatic
	Atlantic Fleet
	Aviation
	Destroyer
	Escort
	Grand Fleet
	Mine Laying

Asia and Europe clasps mounted on the Navy Occupation Service Medal.

Gold, silver and bronze Wintered Over clasps mounted on the Antarctic Service Medal for display purposes.

Continued on page 74

Chief Machinist Mate Michael Ginns with Mobile Base service clasp on his WW I Victory Medal, circa 1920. Naval Historical Center.

Mine Sweeping
Mobile Base
Naval Battery
Overseas
Patrol
Salvage
Subchaser
Submarine
Transport
West Indies
White Sea

Notes on the table: (1) In addition to the Navy World War I Victory Medal clasps listed above, Navy personnel were authorized to wear 11 Army battle and service clasps (see Chapter II for additional details) and (2) the Antarctic Service medal clasp is awarded in bronze, silver and gold (see Chapter IV for additional details).

Miscellaneous Devices
The following miscellaneous devices for wear on the service ribbons were not covered in the preceding tables:

Service Ribbon	Device
Antarctic Ribbon	Bronze, gold or silver 5/16th inch disk
Navy E Ribbon	Smaller silver "E" with laurel wreath

Notes on table: The smaller silver "E' with laurel wreath represents four awards of the Navy "E" service ribbon.

Bronze Maltese Cross of the AEF mounted on the WW I Victory Medal. The cross could only be displayed with Army or Navy "service" clasps and not an Army battle clasp.

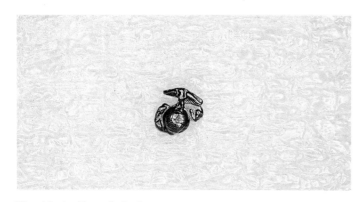

Fleet Marine Force Insignia.

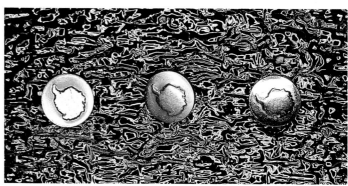

Antarctic service ribbon discs.

9

SERVICE RIBBONS

The ribbons discussed in this Chapter have no corresponding medal. Ribbons worn by Navy personnel measure 3/8th inches by 1 3/8th inches.

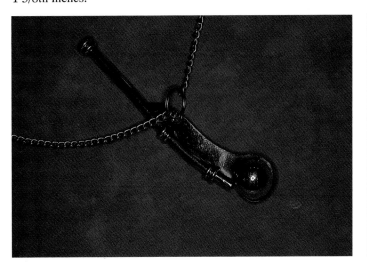

Antique Bosun's Pipe, circa 1900.

Combat Action service ribbon.

Combat Action Ribbon
Established February 17, 1969 retroactive to March 1, 1961, the Combat Action Ribbon is awarded to Navy personnel in the rank of captain or below and Marine Corps personnel in the rank of Colo-nel or below, who have participated in activities which exposed them to enemy fire. The first Navy Combat Action service ribbons were awarded for service during the Vietnam War.

Presidential Unit Commendation
Established on January 10, 1957 the Presidential Unit Commenda-tion (PUC) service ribbon is awarded to units for action which would merit award of the Navy Cross to an individual. Navy personnel are authorized to wear a bronze 3/16th inch star for each subse-quent award and a 3/16th inch silver star in lieu of five bronze stars.

The first PUCs were awarded to the Navy and Marine person-nel involved in the 1941 defense of Wake Island. Commander Wil-liam R. Anderson and the crew of the USS Nautilus (SSN-571) were authorized to wear a gold "N" device on the PUC awarded to them for being the first submarine to sail under the North Pole. Personnel of the USS Triton are authorized to wear a bronze globe for that submarine's submerged circumnavigation of the world.

Presidential Unit Citation service ribbon.

Navy Unit Commendation

Established December 18, 1944, the Navy Unit Commendation ribbon is awarded to any unit, which distinguishes itself in combat against a hostile enemy force under circumstances which would have merited the award of the Silver Star on an individual basis.

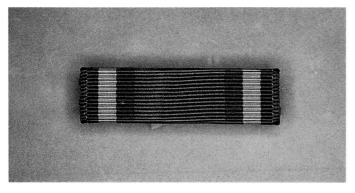

Navy Unit Commendation service ribbon.

Navy Meritorious Unit Commendation service ribbon.

Meritorious Unit Commendation

Established on July 17, 1967, the Meritorious Unit Commendation is awarded to units for valor in combat or non-combat, which would have otherwise merited the award of the Bronze Star on an individual basis.

Navy "E" Ribbon

This ribbon, established on June 1, 1976, replaced a cloth letter "E" which was sewn on the uniform sleeve. The Navy "E" Ribbon is awarded for battle efficiency. Additional silver "E"s are added for subsequent awards. Four or more awards are recognized by wear of a smaller "E" surrounded by a silver wreath.

Navy "E" service ribbon.

Sea Service Deployment service ribbon.

Sea Service Deployment Ribbon

Established on May 22, 1980, retroactive to August 15, 1974, the Sea Service Deployment Ribbon is awarded for 12 months of sea duty or assignment with the Fleet Marine Force for a minimum deployment of 90 days.

Navy Arctic Service Ribbon

Authorized on May 8, 1986, retroactive to January 1, 1982, the Navy Arctic Service Ribbon is awarded for support of the Arctic Warfare Program. To qualify for award of this ribbon, an individual must serve a minimum of 28 days within at least 50 nautical miles of the marginal ice zone (MIZ). The MIZ is defined as an area with at least 10 percent ice concentration. The Navy Arctic Service Ribbon may be awarded only once to a qualifying individual.

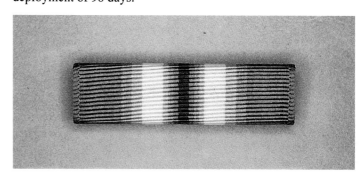

Navy Arctic service ribbon.

Naval Reserve Sea Service Ribbon

Authorized on August 15, 1974, the Naval Reserve Sea Service Ribbon is awarded for a cumulative total of 36 months aboard a Naval Reserve ship. Bronze 3/16th inch stars denote subsequent awards.

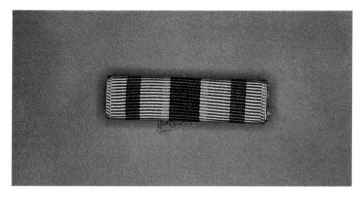

Naval Reserve Sea service ribbon.

Navy & Marine Corps Overseas service ribbon.

Navy & Marine Corps Overseas Service Ribbon

The Naval & Marine Corps Overseas Service Ribbon was authorized on January 1, 1979, retroactive to August 15, 1974. It is awarded for 12 months of overseas duty.

Additional awards are recognized by a 3/16th inch bronze star.

USS Nautilus (SSN-571) shortly after completion of her trip under the North Pole.

10

VARIA

Americans have always received valor medals and other such senior awards from foreign allies. Since World War II however, America's Navy and Marine Corps personnel have been receiving foreign campaign medals in addition to those authorized by the United States. For reference purposes, several of these medals are discussed herein. The practice of the dual campaign medals continues through our most current "campaign" in Bosnia. In several cases, it is a grateful foreign nation, itself, which bestows these medals on U.S. personnel. In other instances, it is an organization such as the United Nations or NATO, which awards the medal.

World War II Soviet Set

While not strictly campaign medals, a grateful Soviet government bestowed the following six medals on 183 U.S. Navy and Coast Guard personnel during World War II. They are presented here as curiosities.

Order of the Patriotic War 1st Class;
Order of the Patriotic War 2nd Class;
Order of the Red Star;
Medal for Valor;
Meritorious Service in Battle Medal; and
the Ushakov Medal.

These medals were awarded to those U.S. crews, which helped supply badly needed war materials to Russian ports during the Soviet struggle against the WW II Nazi German aggressors. The crews not only braved possible U-boat attacks, but also the rugged winter conditions to complete their missions. These crews also qualified for the U.S. ETO Campaign Medal and the World War II Victory Medal.

For those wishing to view these medals, they are on display at the Navy Museum in Washington, D.C.

Soviet Order of the Patriotic War 1st Class.

Soviet Order of the Patriotic War 2nd Class.

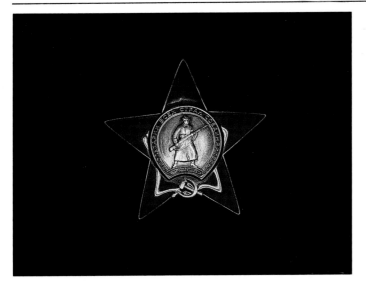

Soviet Order of the Red Star.

Soviet Medal for Valor.

Soviet Meritorious Service in Battle Medal.

Soviet Ushakov Medal.

U.N. Korean Medal

With the Soviet Union boycotting the U.N. General Assembly's Security Council, the North Korean incursion into South Korea was not only condemned, but authorization was given for the establishment of a multinational force to retake South Korea. This would mark the start of a nearly four year war, which would rank as one of the most inconclusive conflicts of all time. When Communist Chinese forces entered on the side of the North Koreans a military stalemate was created. Aligned against the North Koreans and Communist Chinese were forces from the U.S., Belgium, Canada, Denmark, Great Britain, Colombia, Ethiopia, France, Greece, Italy, Luxembourg, the Netherlands, Norway, Philippines, South Korea, Thailand and Turkey.

On December 12, 1950 the General Assembly of the United Nations authorized the award of the U.N. Korean Conflict Medal to all personnel serving in Korea between June 27, 1950 and July 27, 1954 in recognition of their sacrifice.

The bronze medal itself is designed along the lines of traditional British campaign medals with a fixed suspension bar affixed to the planchet. The ribbon features 17 alternating pale blue and white stripes (one for each allied U.N. force). The obverse features a suspension bar with "Korea" as an integral part of the planchet. The reverse of each medal with the exception of those presented to the Commonwealth forces from South Africa is essentially the same. It reads, in each country's native language (12 variations): "For Service in Defence of the Principles of the Charter of the United Nations." U.S. forces were authorized to wear the U.N. medal in conjunction with the Korean Service Medal and the National Defense Service Medal (see Chapter IV). As an aside, South Africans were awarded the medal in silver with a distinctive obverse featuring the intertwined maps of South Africa and Korea. Approximately 800 of these medals were awarded.

United Nations Korean Service Medal obverse.

United Nations Korean Service Medal English language reverse.

Torpedoman First Class Lester L. Osborn, shown with Captain M. F. Leslie, wears the following full size medals (left to right, top to bottom) American Defense Service Medal with Fleet clasp, Asiatic-Pacific Campaign Medal, WW II Victory Medal, China Service Medal, American Campaign Medal and the U.N. Medal for Korean Service circa 1954. National Archives.

Organization of American States Service Medal

This is the medal that never was as far as U.S. forces are concerned.

In 1965, the U.S. Navy again responded to political unrest in the Dominican Republic, which had its roots earlier in the decade. It began when a group of Dominican army officers, loyal to deposed President Juan Bosch, stormed the Presidential Palace on April 24, 1965 in an attempt to overthrow the incumbent government of President Donald Reid Cabral. On April 25, the U.S. 2nd Fleet under the command of Admiral Kleber Masterson began the evacuation of more than 1,200 American citizens. The main elements of the evacuation fleet included the helicopter carrier, Boxer, the Ruchamkin and the Raleigh. On the 27th more than 500 well armed Marines were landed by helicopters from the Boxer to secure the U.S. Embassy. By April 28th, this action had escalated into a full scale U.S. military intervention.

Operation "Barrel Bottom" involved additional Marine and armor landings. They were joined by elements of the U.S. Army's 82nd Airborne Division. More than 23,000 U.S. troops participated in the Dominican Republic intervention. They were eventually joined by an Organization of American States (OAS) multinational peacekeeping force, which established an international security zone. By May 9, the civil war in the Dominican Republic had been brought to an end. The rebels were so stunned by the overwhelming U.S. reaction that they reached a peaceful accord with an interim junta headed by Juan Bosch. U.S. forces began to withdraw soon thereafter. In 1966 Bosch was defeated in free elections and replaced by Joaquin Balaguer, a pro-American moderate.

Many member countries of the OAS, however denounced the U.S.'s unilateral intervention. This subsequently led to the annunciation of the (President Lyndon) Johnson Doctrine, which stated that the U.S. would unilaterally intervene in such matters in order to prevent the establishment of any communist governments (in addition to Cuba) in Latin America.

The U.S. forces participating in the Dominican intervention were authorized to receive the Armed Forces Expeditionary Medal (see Chapter IV) for service in the Dominican Republic and on its contiguous waters between April 28, 1965 and September 28, 1966. These forces would have also received the OAS Medal, except that it was never approved by the Department of Defense. It has been speculated that approval for its issue to American forces was withheld because of the political storm which surrounded the initial U.S. intervention. Original strikes of the medal are available to collectors as curiosity pieces. The OAS medal, itself, is attractively designed and somewhat reminiscent of the U.N. peacekeeping medals. Although many were actually issued to U.S. military personnel, it is still possible to find mint examples of this particular medal.

Among the ships which participated in the 1965 Dominican intervention were:

Boxer
Fort Snelling
Raleigh
Ruchamkin
Wood County

Organization of American States (OAS) Service Medal obverse.

OAS Service Medal reverse.

Republic of Vietnam Campaign Medal

This award was presented by the Republic of Vietnam to all allied forces (U.S., Australia, New Zealand, South Korea, Philippines, Thailand and the Republic of China), who served a minimum of six months or were wounded or killed in action while participating in the military campaigns in Vietnam. The central disk features a map of Vietnam with three red flames representing Vietnam's three main regions. A silver bar on the suspension ribbon reads: "1960 -." No final date was ever added. The Republic of Vietnam ceased to exist on April 30, 1975. U.S. personnel were awarded this medal in addition to the Vietnam Service Medal and the National Defense Service Medal (see Chapter IV). Due to the extensiveness of its award, there are numerous U.S. manufactured versions of this medal in addition to those made by South Vietnam. Japanese and Korean copies also exist.

Republic of Vietnam Campaign Medal obverse.

Republic of Vietnam Campaign Medal reverse.

Kuwait Liberation of Kuwait

This medal was awarded by the government of Kuwait to U.S. and allied personnel who participated in Operations Desert Shield and Desert Storm to liberate Kuwait from the Iraqis occupation forces for one day only during the period August 2, 1990 to August 31, 1993 is required. It was originally designed with a gilt finish, but modified to bronze before being accepted by the Department of Defense for award to U.S. personnel. The planchet, which is linked to a British style fixed suspension bar by a wreath, was modified

Kuwait Liberation of Kuwait Medal (U.S. version) obverse.

Kuwait Liberation of Kuwait (U.S. version) close-up of planchet.

from its original gilt version prior to acceptance by the Department of Defense for award to U.S. personnel. The revised medal has a bronze finish consistent with other domestic campaign medals. U.S. personnel who received this medal were also eligible to receive the Southwest Asia Service Medal and the National Defense Service Medal (see Chapter IV).

Kuwait Liberation of Kuwait Medal original gilt version.

Saudi Arabia Liberation of Kuwait

This medal was awarded by the Kingdom of Saudi Arabia to allied forces serving during Operation Desert Storm from January 17 to February 28, 1991, which resulted in the liberation of Kuwait from the occupying Iraqi forces. U.S. personnel who received this medal were also eligible to receive the Southwest Asia Service Medal and the National Defense Service Medal (see Chapter IV). A palm tree device is authorized for display on the service ribbon.

Saudi Arabia Liberation of Kuwait Medal.

Saudi Arabia Liberation of Kuwait Medal close-up of planchet.

NATO Bosnia Service Medal

The North Atlantic Treaty Organization awards this medal to U.S. and non-U.S. peace keeping forces serving as part of the multinational Implementation Force (IFOR) in the former Republic of Yugoslavia. The IFOR was established as part of the peace agreement signed in Paris by the Presidents of Bosnia-Herzegovina, Croatia and Serbia on December 14, 1995. This agreement ended four years of warfare in the former Republic of Yugoslavia. The role of the IFOR is to keep the former belligerent factions at bay and to pro-tect members of the international war crimes investigation team. The requirements for award of this medal are either 30 days of continuous service in the former Republic or on board ships stationed in the Adriatic Sea or 90 days of service in operational areas outside of the Republic between July 1, 1992 and a date yet to be determined. U.S. personnel must have served under NATO command to qualify.

The medal for non-U.S. IFOR personnel is identical except that it bears a bronze ribbon clasp, which reads: "Former Yugoslavia."

Continued on page 80

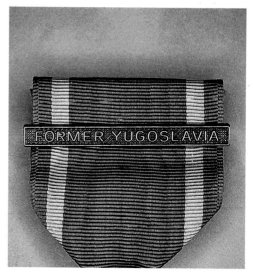

North Atlantic Treaty Organization (NATO) Medal for Bosnia obverse.

NATO Bosnia Medal reverse.

NATO Bosnia Medal clasp "Former Yugoslavia" for non-U.S. participants.

Multinational Force and Observers Medal

Established in 1982, this medal is presented to U.S. and allied personnel serving for six or more months with the Multinational Force and Observers (MFO) on the Sinai Peninsula. The MFO, itself, was established in 1979 and first posted on the Peninsula in August 1981. It is comprised of a maximum strength of 2,600 personnel from 11 nations. Its role is to maintain territorial integrity and stability in this volatile border area between Egypt and Israel following the 1979 Egypt-Israeli Peace Treaty, negotiated at Camp David, Maryland. In addition to monitoring troop strength and movements of the Egyptian and Israeli forces in the area, it is also tasked with protecting the freedom of navigation through the Straits of Tiran. The Straits of Tiran lie at the mouth of the Gulf of Aqaba and provide the Israeli Port of Eilat with access to the Red Sea.

The MFO operates in a fashion similar to that of the United Nations Disengagement Observer Force (UNDOF) on the Golan Heights, but is a separate organization. U.S. participation in the MFO is drawn mainly from the Army's 82nd and 101st Airborne Divisions and the 197th Infantry Brigade. Members of the MFO generally serve three month rotating tours on the Sinai. Technical support, as required, is provided from both other service branches as well as civilians. To date only one U.S. Navy radioman has been awarded this medal. MFO personell wear distinctive orange berets (in contrast to the U.N.'s light blue berets) with a distinctive MFO beret badge, reflecting the stylized dove design of the medal's obverse.

The medal's planchet is attached to a fixed suspension bar similar to that of the U.N. Korean Medal. The ribbon bears the distinctive colors traditionally associated with the MFO.

Multinational Force & Observers Medal obverse.

Multinational Force & Observers Medal reverse.

US Navy River Monitor attacks Viet Cong positions in the Rung Sat Special Zone with its flame throwers, circa 1969. Courtesy John Laemmar, National Vietnam Veterans Art Museum, Chicago.

BIBLIOGRAPHY

Above and Beyond by the editors of Boston Publishing Company. Boston: Boston Publishing Company, 1985.

Conboy, Ken, Bowra, Ken and McCouaig, Simon, *The NVA and Viet Cong.* London, England: Osprey Publishing Ltd, 1991.

Holland, Clive, *Farthest North.* New York: Carroll & Graf Publishers, Inc., 1994.

Hooton, E. R., *The Greatest Tumult The Chinese Civil War 1936-49.* London, England: Brassey's, 1991.

Love, Robert W. Jr., *History of the U.S. Navy:Volume One 1775-1941.* Harrisburg, Pa: Stackpole Books, 1992.

Medals. Hertfordshire, England: Wordsworth Editions Ltd., 1993.

Rodgers, Eugene, *Beyond the Barrier.* Annapolis, Maryland: United States Naval Institute Press, 1990.

Strandberg, John and Bender, Roger James, *The Call of Duty.* San Jose, California: R. James Bender Publishing, 1994.

Sylvester, Jr., John and Foster, Frank, *The Decorations and Medals of the Republic of Vietnam.* Fountain Inn, SC: MOA Press, 1995.

The Military Ribbons of the United States Army, Navy, Marines, Air Force and Coast Guard. Fountain Inn, SC: MOA Press, 1995.

Thomas, Nigel, Abbott, Peter and Chappell, Mike, *The Korean War 1950-53.* London, England: Osprey Publishing Ltd, 1992.

Vernon, Sydney B., *Collectors' Guide to Orders Medals & Decorations* 3rd Edition. Temecula, CA: Sydney B. Vernon, 1995.

Weaver, B., Gleim, A. and Farek, D., *The West Indies Naval Campaign of 1898: The Sampson Medal, The Ships and The Men.* Arlington, VA:1986.

VALUE GUIDE

This Value Guide is provided as a general reference to the reader. In many cases, in particular the commemorative medals, the number of permutations, which impact a medal's value, are far too numerous to cover in a guide such as this. In other cases, the presence of a particular clasp, such as the Wake Island clasp, or an original Silver Citation Star creates a significant impact on the value of the underlying medal. In a 1997 auction guide, an original Wake Island clasp was listed with a suggested bidding price range of $800 to $850 and an original "1919-1920" clasp for the Haitian Campaign Medal was listed with a suggested bidding price range of $700 to $750.

No where is the variation in values more prevalent than in the case of the West Indies/Sampson Commemorative Medal. It would be impossible to cover the values of all of the various permutations which effect the value of this particular medal within the confines of this book. Interested readers would be wise to consult works which concentrate solely on this commemorative medal.

In still other cases, such as the Arctic/Antarctic medals, the rarity of the medals requires that their true value be established via the auction process much like that of a rare painting. To a lesser degree, naming and rim numbers also impact the value of a medal, especially when accompanied by award documents tied directly to the corresponding rim number.

I have tried to present general value ranges for **original** strikes of each medal. Keep in mind that (official) restrikes of the same medal will command significantly less value on the market. The collector should be advised to use this only as a general guide and not a definitive statement as to the value of any particular piece being considered for purchase. In certain cases, where a piece has been struck in such limited numbers that a general value cannot be provided, the code "R" for rare has been substituted for a general value range.

The following miscellaneous codes are also utilized:

FO	Foreign
MC	Marine Corps Version
MN	"M. No." prefixed number
NA	Named
NO	Plain Numbered (no prefix)
SBO	Ship Bar Only

SB/C	Ship Bar with Campaign Bar
SP	Specimen
USN	Navy Version
UN	Unnumbered

The position codes are:

T	Top
B	Bottom
L	Left
C	Center
R	Right
TL	Top Left
TR	Top Right
BL	Bottom Left
BR	Bottom Right

The Value Codes are:

A	Under $100
B	$100 up to $250
C	$251 up to $500
D	$501 up to $1,000
E	$1,000 up to $2,500
F	$2,501 up to $5,000
G	$5,001 up to $10,000
R	Too Rare to Value

Campaign Medals

7	USN	NO	D
7	MC	NO	G
9	USN	NO	D
9	MC	NO	E
10	USN	NO	C
10	MC	NO	E
11	USN	NO	C
11	MC	NO	D
12	USN - 1900	NO	D
12	USN - 1901	NO	E

No.				Code
12		MC	NO	E
14		USN	NO	C
14		MC	NO	D
15		USN	NO	D
15		MC	NO	E
16		USN	NO	B
16		USN	MN	C
16		MC	NO	C
18	TL	USN	NO	C
18	TL	MC	NO	C
18	C	USN w/o clasp 1919-1920	NO	C
18	C	USN w/clasp 1919-1920	NO	E
19		USN	NO	C
19		MC	NO	C
22			UN	A
23			UN	A
24		USN	UN	B
24		MC	NO	C
25		MC w/o clasp	MN	B
25		MC w/clasp	MN	E
26		USN	MN	B
26		MC	MN	B
28		USN	MN	B
28		MC	MN	B
30		USN w/o clasp	UN	B
30		USN w/clasp	UN	E
33		USN	UN	B
33		MC	UN	C
35			UN	A
36			UN	A
37	T		UN	A
37	B		UN	A
38			UN	A
39	T	USN	UN	A
39	T	MC	UN	A
39	B		UN	A
40			UN	A
41	T		UN	A
41	B		UN	A
42			UN	A
43			UN	A
44			UN	A
45			UN	A
46	T		UN	A
46	B		UN	A
47	T		UN	A
47	B		UN	A

Commemorative Medals

No.				Code
49		USN	NA	E
49		USN	UN	C
49		MC	NA	F
51		USN - SBO	NA	C
51		USN - SB/C	NA	D
51		MC - SBO	NA	D
51		MC - SB/C	NA	E
52			NA	G
52			SP	F
53	T			R
53	B			R
54	T			F
54	B			R

Polar Medals

No.			Code
57		NA	R
58		NA	R
59		NA	R
60		NA	R
61	T	NA	C
61	BL	NA	R

Medal Devices

No.		Code
67	TL	D
67	TR	D
67	BL	B, B, B, A
67	BR	B
68	TL	B
68	TR	A, A
68	BL	A

Foreign Medals

No.				Code
74	BL	FO	NO	C
74	BR	FO	NO	B
75	TL	FO	NO	A
75	TR	FO	NO	A
75	BL	FO	NO	A
75	BR	FO	NO	C
76		FO	UN	A
77		FO	UN	A
78	T	FO	UN	A
78	B	FO	UN	A
79	TR	FO	UN	A
79	B	FO	UN	A
80	T	FO	UN	A
80	B	FO	UN	A

All other values, if not separately listed, are Code "A."